No Grid Survival Project

Master the Art of No Grid Living

Alexander Freeman

All rights reserved. No part of this publication may be reproduced, distributed or transmitted in any form or by any means, including photocopying, recording or other electronic or mechanical methods, without the prior written permission of the publisher, except in the case of brief quotations embodied in critical reviews and certain other noncommercial uses permitted by copyright law. For permission requests, write to the publisher.

This book is a comprehensive guide intended for educational and informational purposes only. The publisher and the author make no representations or warranties with respect to the accuracy or completeness of the contents of this book and specifically disclaim all warranties, including without limitation warranties of fitness for a particular purpose. The advice and strategies contained herein may not be suitable for every situation. The publisher is not engaged in rendering professional services and you should consult a professional where appropriate.

The author and publisher shall not be liable for any loss of profit or any other commercial damages, including but not limited to special, incidental, consequential or other damages.
The use of techniques, ideas and suggestions in this book is at the reader's discretion and risk. The publisher and author disclaim any liability, loss or risk, personal or otherwise, which is incurred as a consequence, directly or indirectly, of the use and application of any of the contents of this book.

This publication is designed to provide accurate and authoritative information in regard to the subject matter covered. It is sold with the understanding that the publisher and author are not engaged in rendering legal, accounting or other professional service.

Introduction
Presenting the Work

———— ✧✦✧ ————

In today's world, where our lives are deeply intertwined with modern infrastructure, the idea of living independently from these systems may seem daunting. However, the "No Grid Survival Project" is designed to transform this concept into a practical and achievable reality. This book serves as a comprehensive manual, not just for survivalists, but for anyone seeking independence from external utilities. Whether your goal is to prepare for emergencies, reduce environmental impact or embrace a more self-sufficient lifestyle, this guide will provide the tools and knowledge you need.

At the core of this book is the belief that self-sufficiency is not merely a response to potential crises, but a proactive approach to life. Taking control of your resources and environment brings a profound sense of empowerment and security. This guide is structured to lead you through every step of the journey, offering practical tools, techniques and projects that will help you build and maintain a self-reliant life.

Each chapter is carefully crafted to equip you with the knowledge and skills needed to thrive without relying on modern infrastructure. From creating your own energy systems to securing a sustainable water supply, growing your food and protecting your home, this book covers all aspects of off-grid living. Whether you're a seasoned prepper or a complete beginner, the insights and clear instructions provided in these pages will help you achieve complete independence.

Book Overview

The "No Grid Survival Project - Master the Art of No Grid Living" is meticulously designed as a definitive guide for anyone aspiring to live independently from modern utilities and infrastructure. The book guides readers from understanding the basic principles of off-grid living to mastering advanced techniques that ensure long-term self-sufficiency. It caters to a wide audience, whether you are a complete beginner exploring the idea of living off the grid or an experienced survivalist looking to refine and expand your skills.

The book starts by exploring the importance of self-sufficiency in today's unpredictable world. In a time when natural disasters, economic instability and other unforeseen events can disrupt daily life, the ability to live independently from external systems is not just desirable but essential. This opening section is meant to shift your mindset from one of dependence on modern conveniences to one of empowerment through self-reliance.

Following this foundational understanding, the book delves into practical, hands-on projects designed to help you build the essential components of an off-grid lifestyle. These projects range from simple DIY tasks, like setting up a basic rainwater collection system, to more complex endeavors, such as designing and installing a solar power system. Each project is presented with clear, step-by-step instructions, complete with illustrations and lists of necessary tools and materials, making them accessible to readers with varying levels of experience.

One of the book's key strengths is its comprehensive coverage of all aspects of off-grid living. It goes beyond the basics to explore advanced topics like long-term food preservation, emergency communications and property defense, ensuring you are fully prepared for any scenario. The book also emphasizes adaptability, offering solutions tailored to different environments, whether you live in a rural area, the suburbs or even an urban setting. This flexibility makes the book a valuable resource, regardless of your geographical location or the specific challenges you might face.

Throughout the book, the focus is on practical knowledge that can be immediately applied. This is not a theoretical guide; it is a hands-on manual that encourages you to start building, growing and creating right away. The projects are designed to be scalable, allowing you to start small and expand as your skills and confidence grow. This approach ensures that the path to self-sufficiency is manageable, with progress that is tangible even for beginners.

The book also addresses the psychological and emotional aspects of living off the grid. It recognizes the mental resilience required to transition from a grid-dependent lifestyle to one of complete independence and offers strategies to develop the mindset necessary for this journey. This holistic approach ensures that you are not only physically prepared but also mentally and emotionally equipped to handle the challenges and rewards of off-grid living.

In summary, this book is a thorough and practical guide that covers all aspects of off-grid living. It is designed to be both a comprehensive reference and a step-by-step manual that you can use to build a self-sufficient lifestyle from the ground up. Whether your goal is to prepare for emergencies, reduce your environmental impact or simply live more independently, this book provides the knowledge, tools and inspiration you need to succeed.

What is No Grid Survival?

No Grid Survival refers to the practice and lifestyle of living independently from centralized utilities and infrastructures such as electricity, water, sewage systems and often even communication networks. At its core, it's about reclaiming control over the essential aspects of life by learning to generate, manage and sustain the resources you need for survival, including energy, water, food and shelter. This lifestyle is not merely a response to potential crises; it's a proactive choice to live more sustainably, reduce dependency and cultivate a closer connection to the natural world.

The concept of No Grid Survival is based on the understanding that the systems we rely on daily are inherently vulnerable. Whether due to natural disasters, economic downturns or other unforeseen events, these systems can fail, leaving us without access to the essentials. By adopting a no grid lifestyle, you prepare yourself to thrive even when the modern conveniences we often take for granted are no longer available. It's about resilience and developing the capacity to sustain yourself and your loved ones, regardless of external circumstances.

The appeal of No Grid Survival extends beyond mere survival during emergencies. It embodies a broader philosophy of self-reliance and sustainability. Living off the grid means not just reacting to the possibility of grid failure but actively choosing to live in a way that minimizes your ecological footprint and maximizes your independence. This lifestyle often involves harnessing renewable energy sources such as solar or wind power, collecting and purifying your own water, growing your own food and managing waste sustainably. It's about creating a self-sustained ecosystem that supports your needs with minimal impact on the environment.

A key element of No Grid Survival is the development of practical skills. This includes everything from building and maintaining alternative energy systems to mastering the art of growing and preserving food and constructing shelters that can withstand the elements. These skills are not just useful in a crisis; they are empowering in everyday life. They foster a sense of accomplishment and confidence that comes from knowing you can provide for yourself and your family, regardless of the situation.

Mental preparation is another crucial aspect of No Grid Survival. Transitioning to a lifestyle that is less dependent on external systems requires a significant shift in mindset. It involves embracing a level of uncertainty and being prepared to face challenges head-on. The process of becoming self-sufficient often involves learning through trial and error, adapting to new situations and developing problem-solving skills. It's about cultivating resilience, not just in terms of physical survival, but in terms of mental and emotional strength.

Adopting a no grid lifestyle also means rethinking how you interact with the world around you. It encourages a deeper awareness of the resources you use and a greater appreciation for the natural environment. By learning to live with less and making the most of what's available to you, you become more attuned to the cycles of nature and more mindful of your impact on the planet. This lifestyle is about more than just surviving; it's about thriving in harmony with the world around you, with a sense of purpose and fulfillment that comes from being truly self-reliant.

In summary, No Grid Survival is a holistic approach to living that emphasizes self-sufficiency, sustainability and resilience. It's about taking control of your life and resources, preparing for the uncertainties of the future and cultivating a lifestyle that is both independent and in tune with the natural world. Whether driven by necessity, a desire for independence or a commitment to sustainability, those who embrace No Grid Survival are choosing a path that offers both practical benefits and profound personal rewards.

Essential Tools and Materials

To successfully embark on your journey toward off-grid living, having the right tools and materials is essential. These items form the backbone of your ability to build, maintain and optimize the systems that will allow you to live independently from modern utilities.

Basic Tools

Hammers, screwdrivers and wrenches: crucial for a wide range of tasks, from simple repairs to constructing larger projects.

Pliers: versatile tools needed for bending metal, gripping and cutting wires.

Cordless drill: an indispensable tool for drilling holes and driving screws, especially useful in areas where electricity might be scarce.

Power tools: circular saws, jigsaws and sanders are necessary for cutting and finishing materials like wood and metal.

Gardening tools: shovels, hoes, rakes and pruners are essential for soil preparation, planting and maintaining crops. A wheelbarrow or garden cart is useful for transporting soil, compost and harvested produce.

Essential Materials

Wood and lumber: fundamental for constructing shelters, raised garden beds and other structures. Treated lumber is preferred for outdoor projects due to its resistance to rot and pests.

Metal and hardware: nails, screws, brackets and hinges are critical for securely assembling structures. Plumbing supplies: PVC pipes, fittings and sealants are needed for creating water collection and purification systems. Water filters and pumps ensure that your water supply is clean and accessible.

Electrical components: solar panels, inverters, charge controllers, batteries, wiring and connectors are essential for generating and managing off-grid power.

Gardening supplies: seeds, compost, mulch and irrigation systems are vital for maintaining a productive garden. Heirloom seeds, which are known for their hardiness and ability to produce viable seeds for future planting, are particularly beneficial for long-term sustainability.

Tips for Purchasing and Storing Tools and Materials

Quality: always prioritize high-quality tools and materials, even if they come at a higher initial cost, as they are more durable and reliable in the long run.

Bulk purchasing: whenever possible, buy in bulk to save money and ensure you have enough supplies on hand for various projects.

Proper storage: organize your tools and materials in a dedicated workspace, keeping them protected from the elements and easily accessible for when you need them.

The journey to no grid survival begins with understanding the importance of self-sufficiency and preparing yourself both mentally and practically. This introductory chapter has laid the foundation by providing an overview of the book's objectives, defining key concepts and outlining the essential tools and materials you'll need. As you move forward, remember that each project and skill you acquire brings you one step closer to a more independent, resilient and fulfilling life. The path to self-sufficiency is a challenging one, but with determination, knowledge and the right tools, you are well-equipped to succeed.

Chapter 1
Survival Psychology

In any survival scenario, the mental and emotional state of the individuals involved is just as important, if not more so, than the physical resources at their disposal. Whether you are preparing for a natural disaster, an extended off-grid lifestyle or any situation where modern conveniences are unavailable, understanding and mastering survival psychology is crucial. This chapter delves into the mental preparation required to face such challenges, exploring how to stay calm under pressure, manage stress and cultivate psychological resilience. The mind is often the first and most significant tool in a survivalist's arsenal and by honing it, you increase your chances of successfully navigating any crisis.

Survival psychology begins with mental preparation. The ability to maintain clarity of thought and calmness in the face of adversity can determine the outcome of any survival situation. However, this mental fortitude doesn't come naturally to everyone. It must be developed through intentional practice and awareness. In this chapter, we'll explore various techniques that help you stay calm, manage stress and use effective strategies to prepare mentally for whatever challenges may arise.

Beyond individual mental preparation, the dynamics within a group can significantly impact survival outcomes. Whether you are with family, friends or a community of like-minded individuals, understanding how to manage relationships, conflicts and communication within a group is essential. We'll discuss how to effectively navigate these group dynamics, ensuring that everyone can work together harmoniously and efficiently during high-pressure situations.

Lastly, we'll focus on building psychological resilience. Survival situations often require prolonged endurance, adaptability and the ability to bounce back from setbacks. Developing resilience through mental exercises, learning to strengthen your mind and cultivating flexibility in your approach can make all the difference when facing long-term challenges. The goal is not just to survive, but to do so with a mindset that supports your well-being and increases your capacity to thrive, even in the most difficult circumstances.

Mental Preparation

Mental preparation is the foundation upon which all survival skills are built. In any survival situation, your ability to stay calm, think clearly and make rational decisions can mean the difference between life and death. This section delves into the strategies and techniques that will help you cultivate a mindset capable of withstanding the pressures and challenges that come with living off the grid or facing a crisis.

The first step in mental preparation is learning how to maintain calm in stressful situations. Panic is one of the greatest threats in a survival scenario because it impairs judgment and leads to hasty, often dangerous, decisions. To combat this, you can develop techniques that train your body and mind to respond to stress with calmness and clarity.

Breathing Exercises: one of the most effective ways to stay calm is through controlled breathing. Deep, rhythmic breathing helps slow down your heart rate, lower blood pressure and reduce the production of stress hormones like cortisol. For example, the 4-7-8 breathing technique, where you inhale for four seconds, hold the breath for seven seconds and exhale slowly for eight seconds, can quickly bring about a state of calm. Practicing this technique regularly can help you control your physiological response to stress, making it easier to remain composed in high-pressure situations.

Progressive Muscle Relaxation (PMR): PMR is a method that involves tensing and then slowly releasing different muscle groups in your body, starting from your toes and working up to your head. This practice not only reduces physical tension but also helps to divert your focus from the stressor to the act of relaxation, breaking the cycle of anxiety.

Grounding Techniques: grounding techniques are designed to bring your attention back to the present moment, helping you to disengage from the spiraling thoughts that can accompany stress. Simple grounding exercises include focusing on the sensations in your body, such as feeling your feet on the ground or the texture of an object in your hand. Another effective grounding technique is the "5-4-3-2-1" method, where you identify five things you can see, four things you can touch, three things you can hear, two things you can smell and one thing you can taste. This sensory awareness helps anchor you in the present, reducing the impact of anxiety.

Cognitive-Behavioral Techniques: cognitive-behavioral techniques involve identifying and challenging negative thought patterns that contribute to stress and anxiety. By reframing these thoughts into more positive or actionable perspectives, you can reduce their impact on your mental state. For instance, instead of thinking,

"I'm going to fail," you might reframe it as, "I have faced challenges before and found a way through." This shift in perspective can be empowering and reduce the feeling of being overwhelmed.

Routine and Structure: establishing a routine, even in a survival situation, can provide a sense of normalcy and control, which are both effective in reducing stress. This might include setting specific times for tasks like gathering supplies, preparing food or resting. A structured routine helps to minimize uncertainty and the chaos that can contribute to anxiety.

Visualization: visualization involves mentally simulating scenarios and practicing your responses to them. This mental rehearsal can prepare your brain to react more effectively when faced with real-life situations. For example, you can visualize yourself successfully finding shelter, lighting a fire or navigating difficult terrain. By repeatedly visualizing these tasks, you create neural pathways that make it easier to perform these actions under stress. Visualization also helps to build confidence, as it reinforces the belief that you can handle challenging situations.

Group Dynamics

Group dynamics play a crucial role in survival situations, particularly when resources are limited and tensions are high. Whether you're surviving with family, friends or a group of strangers, the success of your efforts will largely depend on how well you can work together as a cohesive unit. Understanding and effectively managing group dynamics can make the difference between a group that functions smoothly and one that falls apart under pressure. This section delves into the key aspects of group dynamics, including managing relationships and conflicts, defining roles and responsibilities and fostering effective communication within the group.

Managing Relationships and Conflicts: in any survival scenario, conflicts are almost inevitable. High stress, fatigue and the pressures of the situation can lead to misunderstandings and disagreements. However, if managed properly, these conflicts can be resolved before they escalate into serious issues.

Early Recognition and Open Communication: the first step in managing conflicts is recognizing the signs early. Tension often builds gradually, starting with minor disagreements or unspoken frustrations. It's essential to address these issues as soon as they arise rather than letting them fester. Encourage open communication within the group. Each member should feel comfortable expressing their concerns, frustrations and needs without fear of judgment or retribution. This openness can prevent misunderstandings and reduce the likelihood of conflicts escalating.

Conflict Resolution Strategies: when conflicts do arise, having a clear strategy for resolution is vital. Encourage active listening, where each person involved in the conflict listens to the other's perspective without interruption. This helps to ensure that everyone feels heard and respected, which can defuse tension. Mediation, where a neutral third party helps facilitate the conversation, can also be effective in resolving disputes. The goal is to reach a compromise or agreement that everyone can accept, allowing the group to move forward together.

Emotional Intelligence: developing emotional intelligence within the group can significantly improve relationships and reduce conflicts. Emotional intelligence involves being aware of your own emotions and those of others and using this awareness to guide your interactions. Group members with high emotional intelligence are better equipped to handle stress, manage their emotions and empathize with others, all of which contribute to a more harmonious group dynamic.

Roles and Responsibilities within the Group: clearly defined roles and responsibilities are essential for the smooth operation of any group, particularly in a survival situation. When each person knows their role and what is expected of them, it reduces confusion, ensures that all necessary tasks are completed and helps maintain order within the group.

Assigning Roles Based on Strengths: assign roles based on the individual strengths and skills of each group member. For example, someone with medical training should be in charge of first aid, while someone with experience in navigation might take on the role of pathfinder. Assigning tasks based on skills ensures that each job is done efficiently and correctly, which is critical in a survival situation where mistakes can have serious consequences.

Flexibility in Roles: while it's important to have clearly defined roles, it's equally important to maintain flexibility. Survival situations are unpredictable and circumstances can change rapidly. Group members should be prepared to take on different roles as needed. Cross-training, where individuals learn skills outside their primary responsibilities, can be beneficial. This ensures that if one person is unable to fulfill their role due to injury or another issue, someone else can step in without disrupting the group's operations.

Accountability: each group member should be held accountable for their responsibilities. Accountability helps to ensure that tasks are completed and that everyone is contributing to the group's survival. Regular check-ins or briefings can help maintain accountability, where each person reports on their progress and any challenges they are facing. This not only helps keep everyone on track but also provides an opportunity to address any issues before they become larger problems.

Effective Communication: communication is the cornerstone of successful group dynamics. In a survival situation, where time is often of the essence, clear and concise communication can prevent misunderstandings, coordinate efforts and ensure that everyone is working toward the same goal.

Establishing Communication Protocols: before a crisis arises, establish clear communication protocols. This might include regular meetings to discuss plans and progress, as well as emergency communication signals for when immediate action is needed. Having a set structure for communication helps prevent chaos and ensures that everyone is informed and on the same page.

Non-Verbal Communication: in some situations, verbal communication might not be possible, such as when the group needs to remain silent to avoid detection or during noisy environments where verbal instructions could be missed. In these cases, non-verbal communication becomes crucial. Hand signals, visual cues or even pre-agreed body language can be used to convey important messages quickly and discreetly. The group should practice these non-verbal methods regularly to ensure they are understood and effective when needed.

Fostering an Open Environment: create an environment where group members feel comfortable sharing their thoughts, concerns and ideas. This openness is vital in a survival situation, where holding back information or opinions could lead to missed opportunities or dangerous oversights. Encourage a culture of respect, where all voices are heard and decisions are made collaboratively whenever possible. This inclusive approach not only improves communication but also builds trust within the group, which is essential for long-term cooperation.

Psychological Resilience

Psychological resilience is a critical component of survival, enabling individuals to adapt to adversity, recover from setbacks and maintain a positive and proactive mindset in the face of challenges. In survival situations, where physical and mental demands are intense, resilience allows you to endure hardships, stay focused on your goals and ultimately thrive despite the difficulties. This section explores the concept of psychological resilience, how it can be developed and strengthened and practical exercises that can help you build a resilient mindset.

Building Mental Resilience: mental resilience involves withstanding stress and recovering from difficult situations. It's not an inherent trait but a skill that can be cultivated through intentional practice. This begins with adopting a mindset that views challenges as opportunities for growth rather than as obstacles.

Exposure to Controlled Stressors: one of the most effective ways to build mental resilience is by gradually exposing yourself to controlled stressors. This can be done through simulated survival scenarios, where you practice responding to challenges in a controlled environment. For example, you might practice setting up a camp under adverse weather conditions or navigating unfamiliar terrain with limited resources. These exercises help condition your mind to remain calm and focused under stress, improving your ability to cope with real-life situations.

Positive Thinking and Reframing: resilience is shaped by how you perceive and interpret events. By reframing negative experiences positively, you enhance your ability to cope with adversity. For instance, viewing failures as learning opportunities strengthens your resilience.

Mindfulness and Emotional Regulation: mindfulness practices, like meditation, help manage emotions and keep you present in the moment. This approach allows you to remain clear-headed and focused during high-stress situations, preventing panic and anxiety from overwhelming you.

Cognitive Flexibility Training: cognitive flexibility refers to the ability to adapt your thinking in response to changing circumstances. In survival situations, where conditions can shift rapidly, this skill is invaluable. To build cognitive flexibility, regularly engage in activities that challenge your brain to think in new and creative ways.

Puzzles, strategy games and problem-solving tasks are excellent tools for this purpose. Additionally, learning new skills or hobbies that are outside your comfort zone can also enhance your cognitive flexibility by forcing your brain to adapt to unfamiliar challenges.

Adaptability: adaptability are key components of psychological resilience. Survival situations are often unpredictable, requiring you to adjust your strategies and plans on the fly. The ability to pivot and adapt to new information or changing circumstances is crucial for long-term survival.

Embracing Change: the first step in developing adaptability is embracing change rather than resisting it. Change is inevitable in any survival situation, whether it's a shift in weather conditions, the loss of a critical resource or an unexpected challenge. By accepting that change is a part of the process, you can approach it with a mindset geared toward finding solutions rather than dwelling on the difficulties.

Developing a Growth Mindset: a growth mindset is the belief that your abilities and intelligence can be developed through effort, learning and persistence. This mindset is essential for adaptability, as it encourages you to see challenges as opportunities to grow rather than as threats. By adopting a growth mindset, you become more open to learning new skills, trying different approaches and adapting to changing circumstances.

Practicing Flexibility in Decision-Making: flexibility in decision-making involves being open to new ideas, willing to change your plans and able to consider alternative solutions. In a survival context, this might mean re-evaluating your strategy based on new information, being willing to take a different route if your original path becomes too dangerous or adjusting your goals as circumstances change. Practicing flexibility in everyday life – such as by trying new activities, taking on different roles or solving problems in creative ways – can help build this skill.

Learning from Experience: one of the most effective ways to build adaptability is by learning from your experiences. After each survival scenario or challenging situation, take the time to reflect on what worked, what didn't and what you could do differently next time. This reflection helps you internalize the lessons learned, making it easier to adapt and improve in future situations.

Psychological resilience is the bedrock of long-term survival. By building mental resilience, engaging in exercises to strengthen your mind and cultivating adaptability and flexibility, you develop the capacity to endure and thrive in even the most challenging circumstances. Resilience is not just about bouncing back from setbacks; it's about growing stronger through adversity and maintaining a positive and proactive mindset no matter what challenges arise. As you continue to explore the principles and practices of off-grid living and survival, remember that a resilient mind is your greatest asset, enabling you to face the unknown with confidence and determination.

Survival psychology is the cornerstone of effective off-grid living and survival in any crisis. By focusing on mental preparation, managing group dynamics and building psychological resilience, you equip yourself with the mental tools necessary to face whatever challenges come your way. The journey toward self-sufficiency and survival is as much a mental battle as it is a physical one. By mastering the psychological aspects of survival, you significantly increase your chances of thriving, not just surviving, in any situation. As you continue through this book, remember that every project and skill you develop will be more effective if supported by a strong, resilient mind.

Exercise Chapter 1
Daily Meditation for Mental Fortitude

Objective: to strengthen your mental resilience and clarity of thought, practice this meditation exercise daily. This structured exercise focuses on cultivating calmness, reducing stress and enhancing your ability to remain composed and effective in high-pressure situations. It involves four steps that will guide you through grounding techniques, breath control, visualization and reflection to build mental fortitude.

1. Grounding: begin by finding a quiet place where you can sit or stand comfortably. Close your eyes and bring your awareness to the present moment. Feel the contact your body makes with the ground or chair and take note of the sensations in your feet. This practice anchors you in the here and now, helping to create a foundation of stability.

2. Breath Control: shift your focus to your breathing. Use the 4-7-8 breathing technique; inhale deeply through your nose for four seconds, hold your breath for seven seconds and then exhale slowly through your mouth for eight seconds. Repeat this cycle for five to ten minutes. Controlled breathing reduces the production of stress hormones and helps to regulate your physiological response to anxiety.

3. Visualization: after your breath control, engage in a mental visualization exercise. Picture yourself in a challenging situation that you might realistically face... whether it's navigating a difficult terrain, making critical decisions under pressure or managing a conflict. Visualize yourself handling this situation with calmness, precision and confidence. Imagine each step you would take and focus on the positive outcomes of your actions.

4. Reflection: conclude your meditation by reflecting on your experience. Ask yourself how you felt during the visualization. Did you notice any moments of doubt or anxiety? Use this reflection to identify areas where you can improve your mental resilience. With consistent practice, this reflection will help you to recognize and address mental blocks, leading to greater mental strength over time.

Deliverable: a daily journal entry summarizing your meditation experience. Include reflections on how the exercise affected your mental state, any challenges you encountered during visualization and how these practices are influencing your overall resilience and clarity of thought.

This exercise encourages regular engagement with the core concepts of mental fortitude discussed in the chapter, allowing you to integrate these practices into your daily routine and observe tangible improvements in your mental resilience over time.

Chapter 2
Water Collection and Storage

Water is the most essential element for survival. Without a reliable source of clean, potable water, even the most prepared individuals can quickly find themselves in dire situations. In the context of off-grid living or in emergency scenarios, the ability to collect, purify and store water effectively can make the difference between survival and disaster. In this chapter, we will explore the fundamentals of water collection and storage, focusing on the critical importance of ensuring a safe and consistent water supply.

In a world where we are accustomed to turning on a tap to access clean water, it is easy to overlook the complex systems that deliver this resource to our homes. When living off the grid, you are responsible for every drop you consume, making it crucial to understand the various methods of collecting and storing water. Whether you are harvesting rainwater, tapping into natural sources or purifying what you gather, each method requires careful planning and execution to ensure that your water is safe to drink and sufficient for your needs.

Water collection is not just about quantity; quality is equally important. Contaminated water can carry harmful pathogens and chemicals, leading to serious health risks. Therefore, this chapter will guide you through the process of not only gathering water but also purifying and storing it effectively. You will learn about different sources of water, the benefits of rainwater harvesting and the essential systems needed to keep your water supply clean and safe. By the end of this chapter, you will have a solid understanding of how to manage your water resources, ensuring that you and your loved ones remain hydrated and healthy, no matter the circumstances.

Importance of Water Collection

Water collection is the foundation of self-sufficiency when living off the grid or in any survival scenario. Without a reliable source of potable water, even the most meticulously planned survival strategies can quickly unravel. The importance of securing a consistent supply of clean water cannot be overstated. This section explores why potable water is vital, the various sources from which it can be collected and the significant benefits of implementing effective water collection systems.

The necessity of potable water is one of the first principles of survival. The human body can survive for weeks without food but only a few days without water. Water is essential for hydration, but it also plays a critical role in food preparation, sanitation and even maintaining mental clarity and physical performance. In emergency situations, access to clean water can often become a matter of life and death, making it imperative to plan and secure this resource ahead of time.

When planning for water collection, the first step is identifying potential sources of water. Rainwater is one of the most accessible and renewable resources available. Depending on the climate, rainwater can provide a substantial portion of your water needs, particularly if you have an effective collection and storage system in place. Rainwater harvesting is especially beneficial because it reduces dependence on external sources and is typically free from the chemical treatments often used in municipal water supplies. However, it is essential to consider the local environment, as areas with low rainfall may require supplemental sources or more extensive storage solutions to ensure a sufficient supply.

Natural water sources, such as rivers, streams, lakes and ponds, offer another viable option for water collection. These sources are often more reliable in regions with less predictable rainfall. However, water from these sources may contain contaminants, such as bacteria, viruses or chemicals, requiring purification before it is safe for consumption. The accessibility of these sources also varies greatly depending on your location, with some requiring significant effort or infrastructure to tap into effectively. In rural or wilderness settings, surface water is often more abundant, but it comes with the challenge of ensuring that the water is free from pollutants.

Groundwater, accessed through wells, is a third option for water collection. This source is often more consistent and less prone to contamination than surface water, making it a reliable choice for long-term water supply. However, drilling a well can be a significant investment, both in terms of cost and labor. It also requires careful planning to avoid over-extraction and to ensure the water table remains sustainable over time. In many cases, groundwater may still require some level of purification, particularly if the well is shallow or located near potential sources of contamination, such as agricultural fields or industrial sites.

The benefits of water collection extend beyond mere survival. By harnessing natural water sources, you reduce your dependency on external systems and increase your resilience in the face of disruptions, such as natural disasters or infrastructure failures. Additionally, collecting and using local water resources promotes environmental sustainability by reducing the need for energy-intensive water transportation and treatment processes. Over time, a well-managed water collection system can lead to significant cost savings, as you are no longer reliant on municipal water supplies, which can be expensive and may not always be reliable.

In summary, water collection is a critical aspect of self-reliance and survival. Understanding the importance of potable water, identifying viable sources and recognizing the benefits of effective water collection systems are essential steps in ensuring a secure and sustainable water supply. As you delve deeper into this chapter, you will learn how to design and implement water collection systems that meet your specific needs, ensuring that you have access to clean, safe water no matter the circumstances.

Rainwater Harvesting System

Rainwater harvesting is a cornerstone technique for securing a reliable and sustainable water supply, especially in off-grid living situations or areas where municipal water may not be available. This method involves capturing, directing and storing rainwater for various uses, including drinking, cooking, bathing and irrigation. A well-designed rainwater harvesting system not only provides a renewable water source but also contributes to greater self-sufficiency and resilience.

The first step in setting up a rainwater harvesting system is designing a collection network that efficiently gathers rainwater from available surfaces. Typically, roofs are the primary catchment areas due to their broad surface area and natural inclination to direct water. The size and material of the roof play crucial roles in determining the volume and quality of water that can be harvested. Roofs made from non-toxic materials, such as metal, clay or slate, are preferable as they reduce the risk of contaminating the collected water. The slope and surface area of the roof will also influence the quantity of water that can be captured. For instance, a larger roof area with a steep slope can channel water more effectively into the collection system.

Once rainwater is collected from the roof, it needs to be directed into storage tanks. This is achieved using a system of gutters and downspouts that guide the water from the roof into the storage tanks. Gutters should be made from durable materials, such as PVC or stainless steel, to withstand the elements and prevent rusting. The installation of a slight gradient in the gutters ensures that water flows smoothly toward the downspouts, reducing the likelihood of stagnation and debris buildup. To further enhance the system's efficiency, installing gutter guards or mesh screens can help filter out large debris like leaves and twigs before the water enters the storage tanks.

Filters play a crucial role in maintaining the quality of the harvested rainwater. Placing a first-flush diverter at the beginning of the collection system helps divert the initial flow of rainwater, which may contain dust, bird droppings or other contaminants that accumulate on the roof. This initial runoff is directed away from the storage tank, ensuring that the cleaner water is collected. Additional inline filters or sediment traps can be installed along the downspouts to remove finer particles before the water reaches the storage tanks.

The storage tanks themselves are a vital component of the rainwater harvesting system. These tanks come in various sizes and materials, including plastic, fiberglass and concrete. When choosing a storage tank, it's important to consider the material's durability, the tank's capacity relative to your water needs and its placement on your property. Ideally, the tank should be made from food-grade materials to ensure the water remains safe for consumption. Tanks should be installed in a shaded area or painted a light color to minimize temperature fluctuations that can promote bacterial growth. Additionally, tanks should be sealed to prevent contamination from insects, rodents or other pests.

Maintenance is a key aspect of a successful rainwater harvesting system. Regular cleaning of gutters, downspouts and filters is essential to prevent clogging and maintain water quality. The storage tanks should also be inspected periodically for signs of leaks, corrosion or contamination. Depending on the tank material and water usage, periodic disinfection may be necessary to ensure that the stored water remains potable. This can be done using simple methods, such as adding chlorine or using ultraviolet (UV) purification systems.

Beyond providing a dependable water source, rainwater harvesting offers numerous benefits. It reduces reliance on external water supplies, which can be especially critical during droughts or in areas with unreliable water infrastructure. Harvesting rainwater also contributes to environmental sustainability by reducing runoff that can lead to soil erosion and water pollution. Additionally, using rainwater for non-potable purposes, such as irrigation or flushing toilets, can significantly reduce household water consumption, leading to cost savings and less strain on local water resources.

In conclusion, a well-planned and maintained rainwater harvesting system is an invaluable asset for anyone seeking to live independently of traditional water supplies. By understanding the components of the system – from roof catchment areas and gutters to storage tanks and filtration – individuals can design a system that meets their specific needs and ensures a continuous supply of clean, safe water. As climate patterns change and water becomes an increasingly precious resource, the ability to harvest and store rainwater will be an essential skill for sustainable living.

Water Purification and Storage

Ensuring that the water you collect is safe to drink is as critical as the collection process itself. While rainwater and natural sources like rivers or lakes can provide ample water supplies, these sources often contain contaminants that must be removed before the water is safe for consumption. Effective water purification and storage techniques are vital to maintaining a steady, safe water supply in any off-grid or emergency scenario. This section delves into various purification methods, the importance of proper storage systems and the steps required to regularly test and monitor water quality.

The first line of defense in water purification is boiling, a simple yet highly effective method for killing pathogens such as bacteria, viruses and parasites. Boiling water for at least one minute at a rolling boil (or three minutes at higher altitudes) is sufficient to neutralize most harmful organisms. While this method is reliable, it requires a heat source and can be time-consuming, especially when purifying large quantities of water. Moreover, boiling does not remove chemical contaminants or sediments, so it is often used in conjunction with other purification methods.

Filtration is another key technique in water purification, particularly for removing physical impurities, sediments and some pathogens. Filters can range from simple, portable models designed for individual use to more complex systems suitable for an entire household. Ceramic filters, for example, work by forcing water through tiny pores that trap bacteria and other microorganisms, while activated carbon filters are effective at removing chlorine, volatile organic compounds (VOCs) and some heavy metals. Multi-stage filters combine these technologies to provide a more comprehensive purification process, ensuring that water is free from both biological and chemical contaminants.

Chemical purification methods, such as the use of chlorine or iodine tablets, are particularly useful in emergency situations or when other methods are not feasible. These chemicals are highly effective at killing bacteria and viruses, making water safe to drink. However, they can leave an unpleasant taste and do not remove physical impurities or chemical pollutants. Therefore, chemical treatments are often used as a temporary measure or as a backup when traveling in areas where water quality is uncertain.

Ultraviolet (UV) purification is another advanced method that uses UV light to inactivate pathogens by disrupting their DNA. UV purifiers are effective against bacteria, viruses and protozoa, making them a powerful tool for ensuring water safety. The major limitation of UV systems is that they require electricity or batteries and they do not remove chemical contaminants or particulates. Therefore, UV purification is often combined with filtration to achieve comprehensive water treatment.

Once water has been purified, proper storage is essential to prevent recontamination. Water should be stored in clean, food-grade containers with tightly sealed lids to protect it from bacteria, insects and other potential contaminants. The choice of storage material is important: food-grade plastics, stainless steel or glass are preferred because they do not leach harmful chemicals into the water. Containers should be labeled with the date of storage and it is advisable to rotate the stored water periodically – every six months, for example – to ensure freshness.

Storing water in a cool, dark place is crucial to maintaining its quality. Exposure to sunlight and fluctuating temperatures can promote the growth of algae and bacteria, particularly in plastic containers. Therefore, it's recommended to store water in an environment that is consistently cool and away from direct sunlight. If long-term storage is necessary, adding a small amount of unscented household bleach (about 8 drops per gallon) can help maintain water quality by preventing microbial growth.

Regular testing and monitoring of water quality are vital to ensuring the safety of your stored water. Simple home test kits are available that allow you to check for the presence of common contaminants, such as bacteria, nitrates and chlorine. More comprehensive testing, particularly for chemical contaminants like heavy metals or pesticides, may require sending samples to a laboratory. Monitoring the taste, smell and appearance of your stored water can also provide early warning signs of potential problems.

Water purification and storage are critical components of any survival strategy. By utilizing a combination of purification methods – boiling, filtration, chemical treatments and UV purification – you can ensure that your water supply is safe for consumption. Proper storage techniques, including the use of food-grade containers and regular testing, help maintain the quality of your water over time. These practices not only safeguard your health but also provide peace of mind, knowing that your water supply is reliable and secure, even in the most challenging circumstances.

Mastering the collection, purification and storage of water is a cornerstone of off-grid living and emergency preparedness. By understanding the various sources of water, the benefits of rainwater harvesting and the critical importance of proper purification and storage, you equip yourself with the knowledge necessary to secure a safe and reliable water supply. This chapter has provided the tools and techniques you need to ensure that you and your loved ones have access to clean, potable water, regardless of the circumstances. As you continue to build your off-grid lifestyle, remember that water is your most vital resource and managing it effectively is key to your survival and well-being.

Exercise Chapter 2
DIY Rainwater Harvesting System

Objective: to consolidate your understanding of water collection and storage, this exercise will guide you through the process of designing and implementing a rainwater harvesting system for a scenario of your choice. This plan, structured in four steps, will require you to apply knowledge of system design, component selection and maintenance strategies to ensure an effective and sustainable water collection setup.

1. Scenario Selection: choose an environment where you will implement your rainwater harvesting system. Options might include a rural off-grid cabin, a suburban home looking to reduce water costs or an urban rooftop garden aiming for sustainable irrigation.

2. System Design: based on your chosen scenario, design the rainwater harvesting system, including the roof catchment area, guttering, filtration and storage tanks. Consider factors such as roof material, slope, expected rainfall and water needs. Justify your choices by explaining how each component contributes to the efficiency and effectiveness of the system.

3. Component Selection: select the specific components needed for your system, including materials for gutters, downspouts, first-flush diverters and storage tanks. Discuss the reasons for your selections, considering durability, cost and ease of maintenance. For example, you might choose PVC gutters for their cost-effectiveness and stainless steel tanks for their longevity and resistance to contamination.

4. Maintenance Plan: develop a maintenance plan to ensure the long-term functionality and safety of your rainwater harvesting system. This should include a schedule for cleaning gutters and filters, inspecting tanks for leaks or contamination and any periodic disinfection procedures required to keep the water potable. Discuss the importance of each maintenance task and how it contributes to the system's reliability.

Deliverable: a comprehensive rainwater harvesting system plan document that includes your scenario description, detailed system design, rationale for component selection and a maintenance schedule.

This exercise encourages practical application of the chapter's concepts, allowing you to demonstrate a thorough understanding of rainwater harvesting system implementation in a real-world context.

Chapter 3
Food Production

Producing your own food is a cornerstone of self-sufficiency, particularly for those living off the grid or preparing for various survival scenarios. The ability to grow your own crops and raise small animals not only ensures a consistent and reliable food supply but also empowers you with greater control over your diet and resources. In an era where global supply chains can be disrupted by natural disasters, economic instability or other unforeseen events, the skills to produce your own food are invaluable. This chapter guides you through the essential practices of autonomous food production, focusing on the key aspects of home gardening and small-scale animal husbandry. By mastering these practices, you can enhance your resilience, reduce dependency on external systems and enjoy the myriad benefits of fresh, home-produced food.

The journey toward self-sufficient food production begins with a clear understanding of its advantages. Growing your own food not only provides significant financial savings – since the initial investment in seeds, soil and tools quickly pays off – but also ensures that you are consuming fresh, nutrient-dense produce free from harmful pesticides and synthetic chemicals. Beyond the economic and health benefits, the physical activity involved in gardening, along with the psychological satisfaction of nurturing plants and animals, contributes to overall well-being. Additionally, cultivating your own food fosters a deeper connection to nature, teaching invaluable lessons in patience, resilience and environmental stewardship.

To successfully produce your own food, careful planning of your garden is essential. This involves considering factors such as space, climate and crop selection. In the following sections, we will explore the steps to creating a productive home garden, the techniques required for soil preparation and crop management and the benefits and methods of raising small animals like chickens and rabbits. Each aspect of food production is interconnected, contributing to a holistic and sustainable approach to self-sufficiency.

Introduction to Autonomous Food Production

Producing your own food is a fundamental aspect of achieving self-sufficiency, especially in off-grid living or survival scenarios. The ability to cultivate crops and raise small animals ensures a reliable food source, significantly reducing dependency on external systems. In a world increasingly affected by economic instability, climate change and supply chain disruptions, having the skills and knowledge to produce your own food is both practical and empowering.

One of the primary benefits of home cultivation is the control it provides over the quality of your food. By growing your own fruits, vegetables and herbs, you can avoid the use of pesticides and synthetic chemicals commonly found in commercial agriculture, leading to healthier, more nutritious produce. Furthermore, home gardening allows for the cultivation of a diverse range of plant varieties that may not be available in local markets, enabling you to enjoy a personalized and diverse diet.

In addition to health benefits, home cultivation offers significant economic advantages. The initial costs of setting up a garden – such as purchasing seeds, tools and soil amendments – are quickly offset by the savings on groceries. Over time, your garden becomes a sustainable food source, reducing the need for frequent trips to the store and the associated costs of purchasing fresh produce. As food prices continue to rise, the financial savings from home cultivation can become increasingly substantial.

Planning is the cornerstone of successful food production. Before planting, it's essential to assess your available space, local climate and soil conditions. Whether you are working with a small urban balcony or a large rural plot, understanding these factors will help you design a garden that maximizes productivity. For instance, if space is limited, vertical or container gardening can be effective strategies. In larger spaces, traditional row planting or raised beds might be more appropriate.

Another crucial aspect of planning is selecting the right crops. The best crops for your garden will depend on your climate, the amount of sunlight your garden receives and your personal preferences. It's wise to start with easy-to-grow vegetables like tomatoes, lettuce and beans, which are well-suited to most climates and require relatively low maintenance. As you gain experience, you can expand your garden to include more challenging crops, such as root vegetables, perennials and fruit trees, which may require more specialized care.

Companion planting is another important consideration in garden planning. This practice involves growing different plants together that benefit each other in various ways. For example, planting basil near tomatoes can help repel insects that might otherwise damage the tomato plants. Companion planting can also improve soil health, reduce the need for chemical pesticides and increase overall garden productivity.

Autonomous food production extends beyond plant cultivation. Raising small animals, such as chickens and rabbits, further enhances your self-sufficiency. These animals provide a reliable source of protein and can be managed on a small scale, making them suitable even for those with limited space. The waste produced by these animals can also be composted and used to enrich garden soil, creating a closed-loop system that supports both plant and animal life.

In summary, autonomous food production offers numerous benefits, from improved health and financial savings to increased self-reliance and environmental sustainability. By carefully planning your garden and selecting the right crops, you can create a productive and resilient food system that meets your needs and reduces your dependence on external sources. As you continue to develop your gardening and animal husbandry skills, you will find that these practices not only provide food but also enrich your life by fostering a deeper connection to nature and a greater sense of independence.

Home Garden

Establishing a home garden is a fundamental step toward achieving food autonomy, whether you're living off the grid, preparing for emergencies or simply seeking to reduce your reliance on external food systems. A well-planned and carefully tended garden can provide a steady supply of fresh vegetables, fruits and herbs, ensuring that you and your family have access to nutritious, homegrown produce throughout the year. This section delves into the essential aspects of creating and maintaining a productive home garden, from soil preparation and planting techniques to care, harvesting and storage methods.

The foundation of any successful garden is the soil. Healthy soil is teeming with life, including beneficial microorganisms, earthworms and other organisms that contribute to plant health. To prepare your soil, start by testing its composition and pH levels. Soil tests can reveal vital information about nutrient content, pH balance and the presence of contaminants. Based on the results, you may need to amend your soil with organic matter such as compost, manure or peat moss to improve its structure, fertility and ability to retain moisture. Composting your kitchen scraps, garden waste and animal manure is an excellent way to create nutrient-rich compost that can be used to enrich your garden soil naturally.

Once the soil is prepared, the next step is planting. Timing is critical in gardening, as planting at the right time ensures that your crops will thrive in your local climate. Start by researching the optimal planting times for your region, which will vary depending on your climate zone. Most crops can be planted either as seeds directly into the soil or as seedlings that have been started indoors. Starting seeds indoors allows you to extend the growing season, giving your plants a head start before they are exposed to outdoor conditions. This is especially important for warm-season crops like tomatoes and peppers, which require longer growing periods.

The layout of your garden is also crucial. Group plants with similar water and sunlight needs together and consider using raised beds or rows to improve drainage and soil warmth. Raised beds are particularly beneficial in areas with poor soil, as they allow you to control the soil composition more easily. Additionally, raised beds can reduce the strain of bending over to tend your plants, making gardening more accessible.

Care techniques such as watering, mulching and pest management are essential to maintaining a healthy garden. Watering should be done deeply and infrequently to encourage deep root growth, which helps plants become more drought-resistant. Mulching with organic materials like straw, wood chips or leaves helps retain soil moisture, suppress weeds and improve soil fertility as it decomposes. Pest management in a home garden is best approached using integrated pest management (IPM) techniques, which include practices like crop rotation, encouraging beneficial insects and using natural or organic pesticides only when necessary.

As your plants grow, they will eventually reach the point of maturity and be ready for harvest. Knowing when to harvest each type of crop is essential for maximizing flavor, nutritional value and storage life. For instance, vegetables like tomatoes and peppers should be harvested when fully ripe, while others, such as leafy greens, can be harvested continuously as needed. Proper harvesting techniques, such as using clean, sharp tools to avoid damaging plants, will help ensure that your garden continues to produce throughout the growing season.

Once harvested, storing your produce correctly is vital to preserving its freshness and nutritional value. Different crops have different storage requirements: root vegetables like potatoes and carrots store well in cool, dark environments, while tomatoes and cucumbers are best kept at room temperature. Some fruits and vegetables, such as apples and winter squash, can be stored for months under the right conditions, providing a food supply long after the growing season has ended. Canning, freezing and drying are also effective methods for preserving your garden's bounty for use throughout the year.

By mastering the art of home gardening, you create a reliable food source that can sustain you and your family in any situation. A well-maintained garden not only provides fresh, healthy produce but also connects you to the land and fosters a sense of accomplishment and independence. As you continue to hone your gardening skills, you'll find that the rewards of growing your own food extend far beyond the harvest, enriching your life in countless ways.

Raising Small Animals

Raising small animals, such as chickens and rabbits, is a crucial component of achieving food self-sufficiency, especially when living off the grid or in a survival situation. These animals provide a reliable source of protein through meat, eggs and, in some cases, milk. Additionally, they offer other benefits, such as manure for composting, which can enhance the fertility of your garden. This section delves into the benefits of home rearing, the specifics of raising chickens and rabbits and the essential practices for animal management and care.

The benefits of raising small animals at home extend far beyond the immediate food supply they provide. Chickens, for example, are prolific egg producers, with some breeds laying upwards of 250 eggs per year. This consistent supply of fresh eggs offers an excellent source of protein and essential nutrients like vitamins A, D and B12. Beyond eggs, chickens can also be raised for meat, providing a versatile and nutritious addition to your diet. Rabbits are another excellent option for small-scale meat production. They reproduce quickly, require relatively little space and their meat is lean and rich in protein. Additionally, both chickens and rabbits produce manure that can be composted to create nutrient-rich fertilizer for your garden, contributing to a closed-loop system of food production.

When beginning to raise chickens, it's important to select the right breed based on your specific needs, whether for eggs, meat or both. Dual-purpose breeds, such as Rhode Island Reds and Sussex, are popular for their balanced egg-laying and meat-producing qualities. The setup of your chicken coop is critical; it should be spacious enough, with each bird needing about 3 to 4 square feet in the coop and 8 to 10 square feet in the outdoor run. Ensure the coop is well-ventilated to prevent respiratory issues while being secure enough to protect the flock from predators such as foxes and raccoons. Providing nesting boxes in quiet, dark areas of the coop will encourage hens to lay their eggs there, keeping them safe and easy to collect.

Proper nutrition is essential for keeping chickens healthy and productive. A balanced diet of commercial feed, complemented by kitchen scraps, grains and access to pasture for foraging, supports their overall well-being. Clean water should always be available and maintaining hygiene in the coop is vital to prevent the spread of diseases. Regularly inspect your chickens for signs of illness or parasites and address any health issues promptly to maintain a healthy flock.

Rabbits, similar to chickens, require careful consideration of space, diet and care. Housing rabbits in hutches or cages, either outdoors or in a sheltered area, ensures they are safe from predators and adverse weather. Each rabbit needs at least 4 square feet of space, with their living area kept clean and dry to prevent illness. Their diet should consist primarily of hay, which is crucial for digestion and dental health, supplemented with fresh vegetables and quality pellets. Fresh water is also essential for their health.

Breeding rabbits can be an effective way to increase your livestock. A doe (female rabbit) can produce several litters annually, with each litter containing six to ten kits (baby rabbits). To manage breeding effectively, it's important to track cycles and ensure the doe has adequate recovery time between litters. Kits should stay with their mother for at least eight weeks before being weaned.

Animal management also includes regular health checks and ensuring their safety from predators. Whether raising chickens or rabbits, securing their enclosures is paramount to protect them from threats. This can involve reinforcing coops and hutches with durable wire mesh, installing secure locks and possibly using guardian animals to deter predators.

Raising small animals is not just about food production; it also provides a fulfilling daily routine and a connection to nature. The skills you develop in managing these animals can serve as a foundation for expanding into larger-scale animal husbandry or other homesteading practices.

Successfully raising small animals like chickens and rabbits complements your gardening efforts, forming the backbone of a self-sufficient lifestyle. These practices enable you to thrive independently, regardless of external circumstances. As you continue to refine your animal husbandry skills, you'll find that the rewards extend beyond the immediate benefits of fresh food, enriching your life and deepening your connection to the land.

In conclusion, mastering food production through home gardening and small-scale animal husbandry is a vital step toward self-sufficiency and resilience. By cultivating your own crops and raising animals, you not only secure a reliable food source but also gain valuable skills that promote independence and sustainability. This chapter has provided a comprehensive guide to starting your journey in food production, from planning your garden and choosing the right crops to caring for chickens and rabbits. As you continue to develop these skills, you will find that the benefits extend far beyond the food on your table; they include a deeper connection to nature, greater self-reliance and a more fulfilling lifestyle.

Exercise Chapter 3
Starting Your First Home Garden

Objective: to create a detailed plan for starting your first home garden. This exercise will guide you through selecting the right location, choosing appropriate crops, designing the garden layout and establishing a maintenance routine. By the end of the exercise, you will have a comprehensive plan that is ready to implement.

1. Location Selection: begin by evaluating your available space. Consider factors such as sunlight exposure, soil quality and proximity to water sources. Choose the best spot for your garden, taking into account any potential challenges like poor drainage or limited sunlight. Write a brief description of the location and explain why it is ideal for your garden.

2. Crop Selection: based on your climate, gardening experience and personal preferences, choose a variety of crops to grow in your garden. Include at least three different types of plants (e.g., vegetables, herbs or fruits). Provide a short explanation for each choice, focusing on how these crops will thrive in your specific conditions and meet your dietary needs.

3. Garden Layout: design the layout of your garden, incorporating principles such as companion planting, crop rotation and efficient use of space. Decide whether to use raised beds, containers or traditional ground planting. Draw a simple sketch or create a written description of your garden's layout, including the placement of each crop and any pathways or borders.

4. Maintenance Plan: develop a maintenance schedule that covers essential tasks such as watering, weeding, mulching and pest control. Detail how often you will perform these tasks and any tools or materials you will need. Include any specific considerations for the crops you've chosen, such as particular watering needs or potential pest issues.

Deliverable: a comprehensive garden plan that includes the selected location, chosen crops, detailed layout and a maintenance schedule. This plan will serve as your roadmap to starting and maintaining a successful home garden.

This exercise is designed to help you systematically plan your first garden, ensuring that you are well-prepared to grow a variety of crops and maintain them throughout the season.

Chapter 4
Food Preservation

Food preservation is a vital skill for anyone committed to self-sufficiency, whether living off the grid or simply aiming to reduce waste and ensure a stable food supply year-round. As we navigate the challenges of modern life, the ability to preserve food effectively becomes not just a matter of convenience, but a critical part of sustainable living. This chapter explores the principles of food preservation, examining both time-honored traditional methods and modern techniques, while emphasizing the critical role of food safety in the process.

At its core, food preservation is about extending the shelf life of food by controlling the factors that lead to spoilage. Spoilage is typically caused by microorganisms, such as bacteria, molds and yeasts, as well as by enzymatic activity and oxidation. By mastering the art of food preservation, you can ensure that the bounty of your garden, the meat you raise and other seasonal harvests can be enjoyed long after they are fresh, reducing reliance on external food sources.

The methods used to preserve food have evolved over centuries. Traditional techniques like drying, smoking, fermenting and curing have been employed by cultures worldwide, allowing communities to survive through times of scarcity. These methods are often low-tech, relying on natural processes and requiring minimal energy inputs, making them particularly suitable for those living off the grid. On the other hand, modern methods such as canning, freezing and vacuum sealing offer more convenience and versatility, allowing for the preservation of a broader range of foods with greater precision.

However, no matter which method you choose, food safety remains paramount. Improper preservation can lead to foodborne illnesses, making it essential to follow best practices and guidelines meticulously. This chapter will guide you through the principles of safe and effective food preservation, ensuring that your efforts yield long-lasting, nutritious and safe-to-eat products.

Principles of Food Preservation

Food preservation is an essential practice in maintaining a sustainable and resilient lifestyle, particularly for those living off the grid or preparing for long-term emergencies. The principles of food preservation revolve around extending the shelf life of food while ensuring its safety and nutritional value. Understanding these principles allows you to apply various preservation methods effectively, ensuring that your food supply remains safe and edible over time.

The importance of food preservation cannot be overstated, especially when considering the need for a reliable food supply in times of scarcity or uncertainty. Preservation techniques allow you to store food for extended periods, providing access to essential nutrients even when fresh food is unavailable. This capability is crucial in situations where food supply chains might be disrupted, such as during natural disasters or economic instability or when living in remote areas where frequent access to stores is impractical. Preserving food helps reduce waste, maximize the use of seasonal harvests and create a buffer against lean times, ensuring food security throughout the year.

Food preservation methods have evolved over centuries, adapting to the available technologies and the needs of different cultures. Traditional methods like drying, fermenting and curing have been used for millennia, long before the advent of modern refrigeration and freezing techniques. These methods rely on natural processes – such as removing moisture, creating acidic environments or adding salt – to inhibit the growth of spoilage-causing microorganisms. Drying is one of the oldest preservation methods, effective because it removes moisture from food, preventing bacteria from thriving. Smoking not only dries food but also infuses it with preservative compounds from the smoke, adding flavor while extending shelf life. Fermentation, a process used for foods like sauerkraut and kimchi, creates an acidic environment that preserves food while enhancing its nutritional content.

Modern preservation methods, such as canning, freezing and vacuum sealing, have expanded the possibilities for long-term food storage. Canning, developed in the 19th century, uses heat to kill bacteria and seal food in an airtight environment, allowing it to be stored at room temperature for long periods. Freezing halts bacterial growth by storing food at sub-zero temperatures, a method widely used today for its convenience and effectiveness. Vacuum sealing removes air from packaging, reducing the risk of oxidation and spoilage, thereby extending the food's shelf life.

While modern methods often provide convenience and longer shelf life, traditional techniques remain relevant, particularly in scenarios where access to electricity or modern equipment is limited. By understanding the strengths and limitations of both traditional and modern preservation methods, you can choose the most appropriate techniques for your needs, ensuring a stable and varied food supply.

The primary goal of food preservation is to prevent the growth of harmful microorganisms that can cause foodborne illnesses, making food safety the most critical principle in the process. Effective preservation methods involve controlling factors such as temperature, moisture content, pH levels and exposure to air to inhibit bacterial growth and spoilage. One of the biggest risks in food preservation is botulism, a potentially deadly illness caused by toxins produced by Clostridium botulinum bacteria. These bacteria thrive in low-oxygen environments, which is why proper canning procedures are essential to ensure that all spores are destroyed. This typically involves using a pressure canner for low-acid foods, which raises the temperature above the boiling point of water to effectively kill the bacteria.

In drying and smoking, reducing moisture levels sufficiently is key to preventing microbial growth. Proper drying techniques ensure that food is thoroughly dried without becoming susceptible to mold, while smoking adds an additional layer of protection by coating the food with preservative compounds found in the smoke. Fermentation and curing create environments that are inhospitable to harmful bacteria. For fermentation, maintaining the correct salt concentration and keeping the food submerged in brine are crucial steps to prevent spoilage. Curing involves using salt, nitrates or nitrites to preserve food and prevent the growth of bacteria like Clostridium botulinum.

Regardless of the method used, maintaining cleanliness and hygiene throughout the preservation process is essential to prevent contamination. All equipment and containers must be thoroughly sterilized and preserved foods should be stored in appropriate conditions; cool, dark and dry environments are ideal to maintain their safety and quality over time. By adhering to these principles, you can effectively manage your food resources, reduce waste and ensure a reliable food supply, whether in everyday life or during periods of uncertainty.

Drying and Smoking

Drying and smoking are ancient methods of food preservation that not only extend the shelf life of various foods but also enhance their flavors, creating products that are often considered delicacies. These methods work by reducing the moisture content in foods, thereby inhibiting the growth of bacteria, yeast and molds that would otherwise cause spoilage. Each technique has its own distinct advantages and is suitable for different types of food.

Drying Methods

Drying is one of the simplest and most widely used methods of food preservation. It works by removing moisture from food, which significantly slows down the microbial activity that leads to spoilage. There are several drying methods available, each with its own advantages and appropriate applications.

Air Drying: air drying is the most basic method and involves hanging or spreading food in a well-ventilated area where air circulates freely. This method is particularly effective for herbs, some vegetables and certain fruits. The key to successful air drying is ensuring that the environment is dry and warm enough to facilitate moisture loss without promoting mold growth. For example, herbs like thyme oregano and rosemary can be tied in small bunches and hung upside down in a cool, dry place until they are thoroughly dried.

Sun Drying: sun drying is an extension of air drying but takes advantage of direct sunlight to expedite the process. This method is traditionally used in regions with hot, dry climates, where the sun can effectively dry food without the risk of spoilage. Sun drying is most commonly used for fruits like tomatoes, grapes (to make raisins) and apricots. The food is typically placed on drying racks or trays and covered with a fine mesh to protect it from insects and dust while allowing air circulation.

Oven Drying: for those who do not live in climates conducive to air or sun drying, oven drying provides a controlled environment to remove moisture from food. This method involves placing food on baking sheets in an oven set to a low temperature, usually between 140°F and 200°F (60°C to 90°C), with the oven door slightly ajar to allow moisture to escape. Oven drying is versatile and can be used for a wide range of foods, including fruits, vegetables and even meat for making jerky. The process can take several hours to days, depending on the moisture content and the type of food being dried.

Dehydrators: dehydrators are appliances specifically designed for drying food. They work by circulating warm air around the food, which is placed on trays, at a consistent, controlled temperature. Dehydrators are highly efficient, making them ideal for those who frequently dry food. They are particularly useful for drying fruits, vegetables and meat, offering the advantage of even drying with minimal risk of spoilage. The use of a dehydrator also allows for better retention of the food's color, flavor and nutritional value compared to some traditional methods.

Smoking Techniques

Smoking is a method of food preservation that not only dehydrates food but also infuses it with distinctive flavors. Smoking has been used for centuries to preserve fish, meat and even some cheeses, providing both protection from spoilage and a unique taste profile that is highly prized.

Cold Smoking: cold smoking is a technique where food is exposed to smoke in a temperature-controlled environment, usually below 90°F (32°C). This method does not cook the food; instead, it infuses it with smoke flavor while drying it slowly. Cold smoking is typically used for foods that are cured beforehand, such as salmon, ham and bacon. The process canCold smoking is a technique where food is exposed to smoke in a temperature-controlled environment, usually below 90°F (32°C). This method does not cook the food; instead, it infuses it with smoke flavor while drying it slowly. Cold smoking is typically used for foods that are cured beforehand, such as salmon, ham and bacon. The process can take several days to weeks, depending on the food and the desired level of smokiness.

Hot Smoking: in contrast to cold smoking, hot smoking cooks the food while smoking it, usually at temperatures between 160°F and 225°F (71°C to 107°C). This method is faster and can be used to prepare a variety of foods, including fish, poultry and sausages. Hot smoking not only imparts a smoky flavor but also produces a moist, tender product that is ready to eat straight from the smoker. Foods that are hot smoked are often consumed soon after preparation, although they can be stored for short periods under refrigeration.

Building a Smokehouse: for those serious about smoking, constructing a smokehouse provides the ability to smoke large quantities of food at once. A smokehouse is a structure designed to hold food in a controlled environment where smoke can circulate freely. Smokehouses can range from simple, small-scale setups to more elaborate constructions with multiple racks and precise temperature controls. Building a smokehouse requires planning and knowledge of materials, airflow and heat management, but it can be a rewarding investment for those who regularly smoke large amounts of food.

Both drying and smoking are invaluable techniques for anyone interested in self-sufficiency and food preservation. They not only extend the shelf life of food but also add a depth of flavor that is hard to replicate with other preservation methods. By mastering these techniques, you can enjoy a variety of preserved foods that are both practical and delicious, ensuring that you have access to nourishing and flavorful meals regardless of external circumstances.

Fermentation and Brining Techniques

Fermentation and brining are ancient techniques that have been used for centuries to preserve food, enhance its flavor and improve its nutritional value. These processes rely on the natural action of beneficial bacteria and enzymes to convert sugars into acids, alcohol or other compounds that inhibit the growth of spoilage-causing microorganisms. This section will delve into the principles and methods behind fermentation and brining, offering insights into how you can apply these techniques to preserve a wide range of foods at home.

Fermentation Techniques

Fermentation is a metabolic process where microorganisms such as bacteria, yeast or molds convert organic compounds – typically carbohydrates – into alcohol or organic acids. This transformation not only preserves the food but also enhances its flavor, texture and nutritional profile. Fermented foods are rich in probiotics, beneficial bacteria that support gut health and boost the immune system.

One of the most common forms of fermentation is lactic acid fermentation, where lactic acid bacteria (LAB) convert sugars into lactic acid. This process is responsible for the tangy flavor of fermented vegetables like sauerkraut, kimchi and pickles. To start, vegetables are typically shredded or chopped and then mixed with salt. The salt draws out moisture from the vegetables, creating a brine that covers them. The vegetables are then packed tightly into jars or crocks, ensuring they remain submerged in the brine to create an anaerobic (oxygen-free) environment. Over time, the LAB naturally present on the vegetables begin to ferment the sugars, producing lactic acid, which acts as a natural preservative.

Alcoholic fermentation is another widely practiced technique, particularly in the production of beverages like beer, wine and cider. In this process, yeast converts sugars into alcohol and carbon dioxide. This type of fermentation requires careful control of temperature and sugar levels to ensure a balanced and stable product. Alcoholic fermentation is also used in the production of fermented foods like sourdough bread, where the yeast contributes to the dough's rise and flavor development.

Mold fermentation is less common in home food preservation but is crucial in the production of foods like soy sauce, miso and certain cheeses. In these processes, specific molds are introduced to the food, where they break down complex carbohydrates and proteins, resulting in unique flavors and textures. For example, in miso production, soybeans are fermented with koji mold, which converts the starches in the soybeans into sugars and breaks down proteins into amino acids, creating a savory, umami-rich paste.

Brining Techniques

Brining involves soaking food in a solution of water and salt, often combined with sugar, spices and other flavorings. This method not only preserves food but also enhances its flavor and texture. The salt in the brine draws moisture out of the food through osmosis while simultaneously infusing it with flavor. Brining is commonly used for preserving meats, fish and vegetables.

In wet brining, the food is submerged in a saltwater solution. The concentration of salt in the brine is typically around 5-10%, depending on the type of food and the desired end product. Wet brining is often used for poultry, pork and other meats, helping to keep them moist and flavorful during cooking. The food is left in the brine for a period ranging from a few hours to several days, depending on its size and the recipe's requirements. After brining, the food is usually rinsed to remove excess salt and then cooked or further processed.

Dry brining is a variation where salt and other seasonings are rubbed directly onto the surface of the food, drawing out moisture to create a natural brine that the food then reabsorbs. This method is often used for meats and fish, resulting in a flavorful, well-seasoned product. Dry brining is particularly effective for large cuts of meat, such as whole turkeys or roasts, where it can enhance the flavor and tenderness of the final dish.

For vegetables, pickling is a popular brining method where the food is preserved in a vinegar or saltwater solution. Vinegar pickling, also known as quick pickling, involves submerging the vegetables in a vinegar-based brine, which acts as a preservative due to its acidity. Saltwater pickling, on the other hand, relies on the fermentation process to produce lactic acid, which preserves the vegetables in a similar manner to sauerkraut or kimchi. Pickling is a versatile technique that can be used for a wide variety of vegetables, from cucumbers and carrots to beans and peppers.

Fermentation and brining are time-honored methods of food preservation that offer both practical benefits and culinary delights. By understanding and mastering these techniques, you can not only extend the shelf life of your food but also create dishes with complex flavors and enhanced nutritional value. Whether you're fermenting vegetables for a probiotic-rich side dish or brining meat for a tender, flavorful meal, these techniques are invaluable tools in the quest for food self-sufficiency.

Mastering the art of food preservation is a crucial step toward self-sufficiency and resilience. Whether you are drying fruits, smoking meats, fermenting vegetables or canning your garden's bounty, each method offers a way to extend the shelf life of your food, reduce waste and ensure a reliable food supply. By understanding and applying the principles of food preservation, you can take control of your food resources, reduce dependency on external systems and enjoy the rewards of a well-stocked pantry year-round.

Exercise Chapter 4
Canning Your Own Vegetables

Objective: learn the process of canning vegetables to preserve them safely for long-term storage.

Materials Needed: fresh vegetables (e.g., tomatoes, green beans, carrots), canning jars with lids and bands, large pot (for boiling water), pressure canner or boiling-water canner, canning tools (jar lifter, funnel, bubble remover), salt (optional, for flavor), clean towels.

1. **Prepare Your Workspace**: sterilize your canning jars by boiling them in water for at least 10 minutes; keep them hot until ready to use.

2. **Prepare the Vegetables**: wash, peel, chop or slice vegetables as needed; blanch if required (e.g., for green beans).

3. **Pack the Jars**: use a funnel to pack vegetables into hot jars, leaving about 1 inch of headspace; add a pinch of salt if desired.

4. **Add Liquid and Remove Air Bubbles**: pour boiling water or cooking liquid over vegetables, maintaining headspace; remove air bubbles and wipe jar rims clean.

5. **Seal the Jars**: place lids on jars and screw bands until fingertip-tight; do not overtighten.

6. **Process the Jars:** place jars in the canner; process according to the recommended time for the specific vegetable type and altitude.

7. **Cool and Store**: let jars cool undisturbed for 12-24 hours; check seals and store in a cool, dark place.

Deliverable: write a brief report describing your canning process, challenges encountered and include photos of the final canned vegetables.

This exercise provides practical experience in canning, essential for preserving your garden's produce and ensuring food security year-round.

Chapter 5
Alternative Energy

In a world where energy demands are ever-increasing, yet the reliability of conventional power grids can be uncertain, the importance of alternative energy sources cannot be overstated. Whether you are living off the grid, seeking to reduce your environmental footprint or simply looking to ensure energy independence, harnessing alternative energy is a practical and sustainable solution. This chapter explores various methods of generating power autonomously, focusing on solar, hydroelectric and wind energy systems. Each of these energy sources offers unique benefits and challenges and mastering them can provide you with the energy security needed for a self-sufficient lifestyle.

Energy autonomy is not just about cutting ties with the grid; it's about taking control of your energy future. Traditional power sources, heavily reliant on fossil fuels, are not only finite but also subject to price volatility and geopolitical tensions. By contrast, alternative energy sources such as solar, wind and hydroelectric power are renewable, abundant and increasingly cost-effective. They offer the possibility of clean energy that is not only better for the environment but also provides long-term savings and resilience against power outages. As you consider your options for alternative energy, it's essential to understand both the technological and environmental factors involved in each system, ensuring that you choose the best solution for your specific needs.

This chapter will guide you through the key aspects of setting up and maintaining your own alternative energy systems. From the planning and installation of solar panels to the construction of a wind generator, each section provides detailed instructions and insights to help you achieve energy independence. By the end of this chapter, you will have the knowledge necessary to implement and optimize these systems, contributing to a more sustainable and self-reliant way of living.

Home Solar System

Harnessing solar energy for your home is one of the most effective ways to achieve energy autonomy. A well-designed home solar system can provide a reliable source of electricity, reduce your reliance on the grid and significantly lower your energy costs over time. In this section, we will explore the key components and steps involved in planning, installing and maintaining a home solar system.

Planning and Design

The first step in setting up a home solar system is careful planning and design. This involves assessing your energy needs, understanding the local solar potential and determining the best configuration for your solar panels.

1. **Energy Assessment**: start by calculating your household's energy consumption. Review your electricity bills to determine your average monthly and annual usage. This information will help you decide the size of the solar system needed to meet your energy demands.
2. **Site Evaluation**: the effectiveness of a solar system largely depends on the amount of sunlight your location receives. Conduct a site survey to identify the best location for your solar panels, typically a rooftop or a ground-mounted array. The area should have maximum exposure to sunlight with minimal shading from trees, buildings or other obstructions.
3. **System Sizing**: based on your energy assessment and site evaluation, determine the number of solar panels you need. This calculation will also involve considering factors like panel efficiency, tilt angle and the orientation of the panels relative to the sun. A professional solar installer can help optimize the design to ensure maximum energy production.
4. **Budget and Permits**: solar systems require a significant upfront investment, though they offer long-term savings. Create a budget that includes the cost of panels, inverters, mounting equipment and installation labor. Additionally, research local regulations and obtain any necessary permits before beginning installation.

Solar Panel Installation

Once the planning and design phase is complete, the next step is the actual installation of the solar panels. Proper installation is critical to ensure the safety, efficiency and longevity of your solar system.

1. **Mounting the Panels**: solar panels can be mounted on rooftops or on the ground. Rooftop installations are more common and typically involve securing the panels to a racking system that is attached to your roof. The angle and orientation of the panels should be optimized for maximum sunlight exposure.

Ground-mounted systems offer more flexibility in panel positioning but require more space and additional structural support.
2. **Electrical Connections**: after the panels are mounted, they need to be connected to your home's electrical system. This involves wiring the panels to an inverter, which converts the direct current (DC) electricity generated by the panels into alternating current (AC) electricity used by most household appliances. The inverter is then connected to your home's electrical panel, integrating the solar power into your existing system.
3. **System Integration**: depending on your setup, you may also need to install a battery storage system to store excess energy generated during the day for use at night or during power outages. Additionally, a grid-tied system will require a net meter, which allows you to sell excess electricity back to the grid, potentially earning you credits on your utility bill.

Maintenance and Optimization

Maintaining your solar system is essential for ensuring it operates at peak efficiency over its lifespan, which can exceed 25 years with proper care.

1. **Regular Inspections**: periodically inspect your solar panels for dirt, debris or any physical damage. Clean the panels as needed, especially in areas prone to dust, pollen or bird droppings. Most solar panels require minimal cleaning since rain usually washes away dirt, but in dry climates, manual cleaning may be necessary.
2. **Inverter Maintenance**: the inverter is a critical component of your solar system and should be monitored regularly. Most inverters have a display or monitoring system that indicates the system's performance. Keep an eye on these indicators and schedule professional maintenance if you notice any irregularities.
3. **Monitoring Performance**: use a solar monitoring system to track the performance of your solar panels. This system can provide real-time data on energy production, helping you identify and address any issues that could affect efficiency. Monitoring can also help you optimize energy use within your home, ensuring that you make the most of the solar power generated.
4. **Upgrading and Expanding**: as your energy needs grow or as technology advances, you may consider upgrading your system with more efficient panels or expanding the system by adding additional panels. Regularly reassessing your energy needs and the capabilities of your solar system will help you maintain energy independence.

In conclusion, a home solar system is a powerful tool for achieving energy autonomy. By carefully planning, properly installing and diligently maintaining your system, you can ensure a reliable, sustainable source of energy for years to come. Whether your goal is to reduce energy costs, minimize your environmental impact or gain independence from the grid, solar energy offers a practical and effective solution.

Home Hydroelectric System

A home hydroelectric system is a powerful and sustainable way to generate electricity, especially if you have access to a reliable and flowing water source on your property. Unlike solar or wind energy, which can be intermittent, hydroelectric power can provide a continuous and consistent flow of energy, making it one of the most reliable forms of renewable energy. This section will guide you through the key aspects of assessing water resources, building the hydroelectric system and optimizing its efficiency.

Assessing Water Resources

Before you can build a hydroelectric system, it's crucial to evaluate the water resources available on your property. The viability of a hydroelectric system depends on two primary factors: the flow rate of the water and the head (the vertical distance the water falls).

1. **Flow Rate**: the flow rate is the volume of water that passes a specific point in a stream or river per unit of time, typically measured in liters per second (L/s) or cubic feet per second (cfs). To measure the flow rate, you can use a flow meter or estimate it by timing how long it takes to fill a container of a known volume. The higher the flow rate, the more energy you can generate.
2. **Head**: the head is the vertical drop from the point where the water is diverted to where it powers the turbine. It's a critical factor because the greater the head, the more potential energy is available to convert into electricity. The head can be measured using topographical maps or by direct measurement with tools like a transit level or a simple water level.
3. **Site Selection**: after measuring the flow rate and head, select a site on your property that maximizes these factors. Ideally, the site should also be close to where the electricity will be used to minimize transmission losses. Consider the environmental impact and ensure that the system doesn't significantly alter the natural flow of the water or harm local ecosystems.

Building the System

Once you've assessed your water resources and selected the optimal site, the next step is to build the hydroelectric system. A typical home hydroelectric system includes a water intake, a penstock, a turbine, a generator and electrical controls.

1. **Water Intake**: the water intake is where the water is diverted from the stream or river. It should be designed to allow sufficient water flow while preventing debris and sediment from entering the system. A screen or grate is typically used to filter out large debris. The intake should be positioned in a way that maintains a steady flow of water, even during dry periods.
2. **Penstock**: the penstock is the pipeline that carries water from the intake to the turbine. It should be made of durable materials like PVC, steel or polyethylene and should be sized appropriately to handle the flow rate without significant friction losses. The penstock needs to be securely anchored along its path to prevent movement due to water pressure or environmental factors.

3. **Turbine**: the turbine is the heart of the hydroelectric system. It converts the kinetic energy of the flowing water into mechanical energy. The type of turbine used will depend on the head and flow rate of your site. Common types include Pelton wheels for high head, low flow sites and Kaplan or Francis turbines for low head, high flow sites. The turbine must be carefully matched to the site conditions to ensure optimal performance.
4. **Generator**: the mechanical energy from the turbine is converted into electrical energy by a generator. The generator's size and type should be chosen based on the expected output from the turbine. In small-scale hydroelectric systems, permanent magnet generators are often used due to their efficiency and reliability.
5. **Electrical Controls and Wiring**: the electricity generated by the system is typically in the form of alternating current (AC) or direct current (DC). It needs to be regulated and possibly converted depending on your power needs. This involves using controllers, inverters and safety disconnects. The power can be used directly, stored in batteries or fed into the grid if your system is grid-tied.

Efficiency Optimization

To maximize the efficiency and longevity of your home hydroelectric system, regular maintenance and optimization are essential.

1. **Regular Maintenance**: inspect the intake for debris and clear it regularly to maintain optimal water flow. The penstock should be checked for leaks or blockages and the turbine and generator should be inspected for wear and tear. Proper lubrication of moving parts and checking electrical connections for corrosion are also vital maintenance tasks.
2. **System Monitoring**: use monitoring tools to track the system's performance. This includes measuring the output voltage, current and overall power generation. Monitoring helps identify any inefficiencies or issues that may arise, allowing you to address them promptly.
3. **Seasonal Adjustments**: depending on your water source, you may need to adjust your system seasonally. For instance, during periods of high water flow, you might need to reduce the intake to prevent overwhelming the system. Conversely, in dry seasons, you might need to conserve water to maintain a steady power supply.
4. **Upgrading Components**: as technology advances, consider upgrading your system components to more efficient or durable models. For example, newer turbine designs or more efficient generators can increase the overall output of your system.

A home hydroelectric system offers a reliable and sustainable source of energy that can significantly reduce your dependence on conventional power sources. By carefully assessing your water resources, building the system with precision and maintaining it effectively, you can enjoy the benefits of clean, renewable energy for years to come. This approach not only supports a self-sufficient lifestyle but also contributes to the preservation of natural resources and the environment.

Wind Generator Construction

Constructing a wind generator at home is a rewarding project that provides a sustainable and renewable source of energy. Wind power, harnessed through a wind turbine, converts kinetic energy from wind into electrical energy, which can be used to power various household appliances or charge batteries for later use. This section will guide you through the essential steps in constructing your own wind generator, including the materials and tools required, the construction procedure and tips for installation and maintenance.

Required Materials and Tools

To build a functional wind generator, you will need the following materials and tools:

- **Materials**: alternator (or DC motor), blades (can be made from PVC pipe, wood or pre-fabricated blades), tail assembly (to keep the turbine facing the wind), tower or pole (metal or wood, depending on the height and strength needed), hub (to attach the blades to the alternator), wiring (gauge appropriate for the current), charge controller, batteries (for storing energy), inverter (if you plan to convert DC to AC) and miscellaneous hardware (bolts, nuts, washers and connectors).
- **Tools**: drill with bits, saw (if cutting blades or wood), wrench set, wire strippers, multimeter (for testing), screwdrivers, pliers, measuring tape and a ladder (for installation).

Construction Procedure

1. **Assemble the Alternator and Hub**: the alternator or DC motor is the heart of your wind generator. It converts the rotational energy of the blades into electricity. Begin by securely attaching the hub to the alternator shaft. The hub is where the blades will be mounted. Ensure that the connection is firm to prevent wobbling during operation.
2. **Prepare the Blades**: the blades catch the wind and spin the alternator. Depending on your design, you can either purchase pre-made blades or create them from materials like PVC or wood. If you are making your own, cut the blades to the desired shape and length, ensuring they are balanced and of equal weight to prevent uneven spinning. Sand and smooth the edges to reduce air resistance.
3. **Attach the Blades to the Hub**: securely fasten the blades to the hub using bolts and nuts. Make sure that the blades are evenly spaced and angled correctly to catch the wind efficiently. The typical angle is about 10 to 15 degrees, but this can vary depending on the blade design and wind conditions.
4. **Construct the Tail Assembly**: the tail assembly keeps the turbine facing into the wind. Construct a simple tail vane from lightweight materials like metal or plastic and attach it to the back of the alternator housing. The tail should be large enough to effectively steer the turbine but not so heavy that it affects the balance.
5. **Mount the Turbine on the Tower**: the tower or pole will elevate your wind generator to a height where it can catch the wind more effectively. Attach the alternator and blade assembly to the top of the tower. If using a metal tower, ensure it is grounded to protect against lightning strikes. For wooden towers,

make sure it is treated to withstand weather conditions. The height of the tower will depend on your location and wind patterns, but a typical height is between 20 to 40 feet.
6. **Wiring the System**: connect the wiring from the alternator to the charge controller and then to the batteries for energy storage. The charge controller is essential as it regulates the voltage and current coming from the turbine, preventing overcharging of the batteries. If you plan to use the electricity for AC appliances, connect the batteries to an inverter that converts DC to AC power.
7. **Install a Brake System (Optional)**: some wind generator designs include a brake system that can stop the blades from spinning during storms or when maintenance is needed. This can be a simple manual brake or an electronic control integrated with the system.

Installation and Maintenance

Once your wind generator is built, proper setup and ongoing maintenance are crucial for top performance.

1. **Installation**: choose an open location with minimal obstructions like trees or buildings to maximize wind exposure. The wind generator should be installed at least 20 feet above any nearby obstacles. Secure the base of the tower firmly into the ground, using concrete if necessary, to prevent it from tipping over. Ensure all electrical connections are waterproofed and protected from the elements.
2. **Initial Testing**: before fully connecting the generator to your home system, conduct initial tests to ensure everything is functioning correctly. Check the blade rotation, electrical output and stability of the tower. Use a multimeter to measure the voltage output and ensure it matches your expectations.
3. **Regular Maintenance**: periodically inspect the blades, hub and tail assembly for signs of wear or damage. Lubricate any moving parts as necessary to reduce friction and noise. Check the wiring and electrical connections for corrosion and clean them as needed. Ensure the tower remains stable, especially after severe weather events.
4. **Performance Optimization**: monitor the energy output of your wind generator regularly. If you notice a drop in performance, investigate potential causes such as blade imbalance, obstructions or electrical issues. Adjusting the blade angle or upgrading components like the alternator or blades can improve efficiency.

Building a wind generator requires careful planning, precision and regular upkeep, but the reward is a reliable and renewable energy source that can significantly reduce your dependence on the grid. By following these steps and maintaining your system, you can enjoy the benefits of wind power for years to come, contributing to a more sustainable and energy-independent lifestyle.

Transitioning to alternative energy sources is a crucial step toward energy independence and environmental sustainability. By installing solar, hydro and wind power systems, you can significantly reduce reliance on traditional energy, lower your carbon footprint and ensure a stable, reliable power supply for your home. Each system offers unique benefits and challenges, but with careful planning, proper setup and regular upkeep, they provide long-term energy security, peace of mind and enhanced self-sufficiency for a sustainable future for all.

Exercise Chapter 5
Building a Solar Charger for Small Devices

Objective: learn to build a simple solar charger to power small devices like phones, tablets or rechargeable batteries using readily available materials.

Materials needed: small solar panel (5V–12V output), blocking diode (1N5819 or similar), USB charging circuit (optional), battery pack (for storing energy), wires, soldering iron, solder, multimeter, enclosure box (for housing the components), connectors (USB or other as needed), small screws and tools (screwdriver, wire strippers).

1. Prepare the Solar Panel: connect the solar panel's positive terminal to the blocking diode's anode (non-striped end) and the cathode (striped end) to the circuit's positive wire. This prevents the battery from discharging back into the solar panel.

2. Attach the Charging Circuit: if using a USB charging circuit, connect it to the diode's output. The solar panel's negative wire connects directly to the circuit's input, allowing the circuit to regulate voltage and current.

3. Install the Battery Pack (Optional): connect the battery pack's terminals to the solar panel and diode circuit, ensuring correct polarity. This allows energy storage for later use.

4. Test the Circuit: use a multimeter to check the voltage at different points in the circuit, especially at the output where you will connect your device. Ensure the voltage is within safe limits for your device (typically 5V for USB devices).

5. Mount Components in the Enclosure: place the solar panel, diode, battery pack and charging circuit inside the enclosure box, ensuring the solar panel is securely mounted and exposed to sunlight. The enclosure protects the components from the elements.

6. Final Assembly and Testing: close the enclosure and test the entire setup by connecting a small device to the charger. Ensure it charges effectively and the setup remains stable under various conditions.

Deliverable: write a brief report detailing your solar charger construction process, any challenges faced and how you overcame them. Include photos of your final assembly and discuss potential improvements for efficiency.

This exercise is a hands-on introduction to using solar power for small-scale applications, essential for those interested in off-grid living or reducing reliance on conventional power sources.

Chapter 6
Property Security and Defense

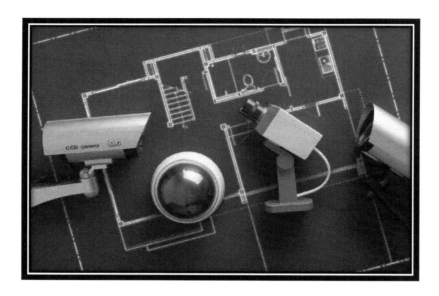

In an increasingly uncertain world, ensuring the security and defense of your property is paramount for maintaining peace of mind and safeguarding your loved ones. Whether you are living in an urban environment, a rural area or off the grid, understanding the principles of property security and implementing effective defense strategies can make a significant difference in deterring potential threats. This chapter delves into the core aspects of home security, from risk assessment and defense planning to the construction of DIY security systems and the development of self-defense techniques. By mastering these skills, you will be better equipped to protect your home and family from various threats, both natural and man-made.

Home security is not just about installing a few locks and hoping for the best. It involves a comprehensive approach that considers the unique vulnerabilities of your property and the potential risks you face. By conducting a thorough risk analysis and planning your defenses accordingly, you can create a layered security system that deters intruders, detects unauthorized access and defends against potential attacks. This chapter will guide you through the process of analyzing risks, planning defenses and implementing practical security measures, all while considering the specific needs and circumstances of your home and lifestyle.

In addition to traditional security measures, this chapter also covers the construction of DIY security systems, including building security fences, installing alarm systems and using surveillance cameras. These systems not only enhance your property's security but also provide you with greater control and flexibility in how you protect your home. Furthermore, we will explore essential self-defense techniques, from basic training to the use of defense tools and emergency planning. Whether you are preparing for a worst-case scenario or simply want to feel more secure in your daily life, this chapter offers the knowledge and skills you need to build a robust property security and defense system.

Introduction to Home Security

Home security plays a critical role in creating a safe living environment. Beyond merely protecting your belongings, a well-secured home contributes significantly to the mental and emotional well-being of its occupants. When you know your home is secure, you can focus on other aspects of your life without the constant worry of intrusions or emergencies. Effective home security serves as both a deterrent and a defense mechanism against potential threats.
A secure home sends a clear message to potential intruders: it is not an easy target. Visible security measures such as surveillance cameras, alarms and sturdy fences significantly reduce the likelihood of a break-in by increasing the perceived risk for the intruder. This not only protects your property but also contributes to a safer community overall, as criminals are often deterred from targeting well-secured neighborhoods.

The first step in achieving comprehensive home security is to conduct a detailed risk analysis. This involves identifying the unique vulnerabilities of your property, understanding the specific risks you face based on your location and assessing the potential impact of those risks. For instance, urban homes might be more susceptible to crimes like burglary, while rural properties could face threats from wildlife or delayed emergency response times. By understanding these risks, you can prioritize your security efforts where they are needed most.

A thorough risk analysis should also consider factors such as the visibility of valuable items, the ease of access to entry points and your daily routines. For example, predictable patterns of leaving and returning home can be exploited by criminals who observe your habits. Additionally, understanding the local crime trends can provide valuable insights into the most common types of threats in your area, allowing you to tailor your security measures accordingly.

Defense planning is the strategic process of developing and implementing measures to protect your home from identified risks. This involves creating a multi-layered security approach that addresses both physical barriers and psychological deterrents. A robust defense plan starts with reinforcing the most vulnerable points of entry, such as doors, windows and garages, using high-quality locks, reinforced door frames and shatter-resistant glass.

Incorporating technology into your defense plan can further enhance your home's security. Alarm systems, motion detectors and surveillance cameras not only help detect and deter intrusions but also provide real-time alerts to you and, if necessary, the authorities. These systems can be integrated into a broader smart home network, allowing you to monitor and control your security remotely.

However, the effectiveness of any defense plan also depends on the people living in the home. Educating all household members on security protocols is essential. Everyone should know how to operate the security systems, understand the importance of keeping entry points secured and be familiar with the procedures to follow in case of a breach. Regular drills and updates to the plan ensure that everyone remains vigilant and prepared.

Finally, defense planning should include emergency preparedness. This involves creating an emergency plan that outlines evacuation routes, safe rooms and communication strategies. In the event of a home invasion, fire or natural disaster, having a clear and practiced plan can be the difference between safety and disaster. Regularly reviewing and updating your defense plan as circumstances change will help maintain the security of your home.

DIY Security Systems

In an era where security is a growing concern, taking a proactive approach to protect your property is essential. DIY security systems offer an effective and customizable solution that allows you to tailor the protection measures specifically to your home's needs. This section explores the key components of DIY security systems, including building security fences, installing alarm systems and utilizing surveillance cameras. By combining these elements, you can create a robust security network that deters intruders, detects unauthorized access and provides peace of mind.

Building Security Fences

A security fence serves as the first line of defense, marking the boundary of your property and creating a physical barrier that deters unauthorized entry. When constructing a security fence, consider the following key factors: materials, height and additional deterrents.

Materials: the choice of materials for your security fence depends on your specific needs and aesthetic preferences. Metal fences, such as those made from wrought iron or steel, are highly durable and difficult to breach, making them ideal for security purposes. They can be designed with pointed tops or spikes to prevent climbing. Wooden fences, while more visually appealing, can be reinforced with metal plates or wires to enhance their security. Chain-link fences are a popular option due to their affordability and visibility, allowing you to see through them while still providing a physical barrier.

Height: a security fence should be tall enough to discourage climbing, typically at least 6 to 8 feet high. For added security, the top of the fence can be angled outward or topped with barbed wire or razor wire, though these options may be subject to local regulations.

Additional Deterrents: to further increase the effectiveness of your security fence, consider integrating additional deterrents. Lighting along the perimeter of the fence eliminates shadows and dark areas, making it more difficult for intruders to approach undetected. Motion-activated lights are particularly effective, as they startle potential intruders and increase the likelihood of them being spotted. Planting thorny bushes or hedges along the inside of the fence can also serve as a natural deterrent, making it more difficult to climb over.

Installing Alarm Systems

An alarm system is a critical component of any home security setup. It serves as both a deterrent and an early warning system, alerting you to potential intrusions. When installing an alarm system, several factors should be considered:

- **System Selection**: there are various types of alarm systems available, ranging from basic door and window sensors to more advanced systems that include motion detectors, glass break sensors and smart home integration. Choose a system that matches the specific needs of your home. For example, if your property is particularly large or has multiple access points, you might opt for a system with comprehensive coverage, including both indoor and outdoor sensors.
- **Placement of Sensors**: the effectiveness of an alarm system largely depends on the strategic placement of its sensors. Door and window sensors should be installed on all potential entry points, including less obvious ones like basement windows or garage doors. Motion detectors are best placed in high-traffic areas inside the home, such as hallways, living rooms and near staircases. If you have pets, consider using pet-friendly motion sensors that are less likely to trigger false alarms.
- **Connection and Monitoring**: modern alarm systems often come with options for professional monitoring services, which can alert authorities in case of an intrusion. However, if you prefer a fully DIY approach, many systems offer self-monitoring via smartphone apps, allowing you to receive alerts and control your system remotely. Ensure that your system is connected to a reliable power source and has battery backup in case of power outages.
- **Testing and Maintenance**: regularly testing your alarm system is essential to ensure that all sensors and components are functioning correctly. Replace batteries as needed and update software if your system is connected to a smart home network. Routine maintenance can prevent potential malfunctions and ensure that your alarm system remains effective over time.

Using Surveillance Cameras

Surveillance cameras are a powerful tool in any home security system, providing real-time monitoring and recording of activities around your property. They can deter criminals by increasing the perceived risk of being caught and can also provide valuable evidence in the event of a security breach.

Camera Selection: choose cameras that meet your specific security needs. Options include indoor and outdoor cameras, wired or wireless systems and cameras with features like night vision, motion detection and high-definition video. Outdoor cameras should be weatherproof and have a wide field of view to cover large areas like driveways and backyards. Indoor cameras are ideal for monitoring entry points, hallways and other critical areas inside the home.

Placement and Coverage: the placement of surveillance cameras is crucial for maximizing their effectiveness. Outdoor cameras should be positioned to cover all potential entry points, including doors, windows and gates. Consider installing cameras at an elevated height to avoid tampering and angle them downward for a clear view of the area. Indoor cameras should cover key areas like entryways, hallways and rooms with valuable items. It's important to ensure that there are no blind spots where intruders could move undetected.

Integration with Other Security Systems: for enhanced security, integrate your surveillance cameras with other components of your DIY security system. Many modern cameras can be connected to alarm systems, allowing them to start recording as soon as a sensor is triggered. Additionally, integrating cameras with a smart home system enables you to monitor your property remotely, receive alerts and even communicate through built-in microphones and speakers.

Recording and Storage: decide how you will store the footage captured by your surveillance cameras. Options include cloud storage, which provides easy access and security and local storage on a digital video recorder (DVR) or network video recorder (NVR). Consider the retention period for your recordings based on your storage capacity and security needs. Regularly check that your cameras are recording correctly and that storage devices are functioning properly.

In conclusion, building a comprehensive DIY security system involves combining physical barriers like security fences with technological measures such as alarm systems and surveillance cameras. Each component plays a vital role in creating a secure home environment, offering protection, deterrence and peace of mind. By taking a proactive approach to home security, you can safeguard your property and ensure the safety of your loved ones.

Self-Defense Techniques

In an unpredictable world, the ability to protect yourself and your loved ones is an invaluable skill. Self-defense techniques not only empower you physically but also mentally, enhancing your confidence and preparedness in potentially dangerous situations. This section delves into the essential components of self-defense, including basic training, the use of defense tools and emergency planning. Each aspect plays a crucial role in creating a comprehensive self-defense strategy that can be adapted to various scenarios, from home invasions to personal safety in public spaces.

Basic Training

Basic self-defense training lays the foundation for effectively responding to threats. It involves learning physical techniques, understanding situational awareness and developing the mental fortitude to stay calm under pressure. The goal of basic training is not just to teach you how to fight but to equip you with the skills to avoid confrontation whenever possible and to defend yourself effectively when necessary.

Situational Awareness: one of the most critical aspects of self-defense is situational awareness. This means being aware of your surroundings at all times and recognizing potential threats before they escalate. Situational awareness involves using all your senses to monitor your environment, identifying exits and noting any unusual behavior. By staying alert and conscious of your surroundings, you can often avoid dangerous situations before they develop.

Physical Techniques: self-defense techniques focus on neutralizing threats with minimal force. Basic moves include strikes (such as punches, kicks and knee strikes), blocks (to deflect incoming attacks) and escapes (such as breaking free from grabs or holds). Training should emphasize the use of leverage, speed and precision rather than brute strength, making these techniques accessible to people of all sizes and strengths. Commonly taught moves include palm strikes, elbow strikes and knee strikes, which target vulnerable areas of an attacker's body, such as the eyes, throat and groin.

Practice and Repetition: mastery of self-defense techniques comes through practice and repetition. Regular training helps develop muscle memory, ensuring that your responses to threats are quick and instinctive. Practice can be done alone through drills or with a partner in controlled sparring sessions. Joining a self-defense class or martial arts program can provide structured training and access to experienced instructors.

Using Defense Tools

While physical self-defense techniques are essential, the use of defense tools can provide an additional layer of protection. Defense tools are designed to be easily carried and quickly deployed in emergencies, giving you an advantage in a confrontation.

Pepper Spray: pepper spray is one of the most commonly used self-defense tools. It is non-lethal and can incapacitate an attacker temporarily by causing intense burning and irritation to the eyes, skin and respiratory system. Pepper spray is compact, easy to carry and can be effective at a distance of several feet, allowing you to create a safe escape route. When using pepper spray, aim for the attacker's face, specifically the eyes and be prepared to move quickly to a safe location immediately after deployment.

Personal Alarms: personal alarms are small, portable devices that emit a loud, piercing sound when activated. The purpose of a personal alarm is to startle the attacker and attract attention to the situation, potentially deterring the attacker and drawing help from nearby people. These devices are easy to carry and can be attached to keychains or bags for quick access.

Tactical Flashlights: tactical flashlights serve a dual purpose in self-defense. They can be used to disorient an attacker with a bright, blinding light, giving you a few crucial moments to escape or prepare for defense. Additionally, many tactical flashlights are designed with a hardened bezel that can be used as an impact weapon in close combat situations.

Stun Guns and Tasers: stun guns and tasers are electroshock devices designed to incapacitate an attacker by delivering a high-voltage electrical charge. Stun guns require direct contact with the attacker, while tasers can be used from a distance, firing barbed electrodes that transmit the charge. These devices can temporarily paralyze an attacker, giving you time to escape. However, they require close proximity and some familiarity with their operation to be used effectively.

Emergency Planning

Emergency planning is a critical aspect of self-defense that involves preparing for various scenarios where your safety could be at risk. Having a well-thought-out plan in place can significantly increase your chances of staying safe during an emergency.

Home Defense Plan: developing a home defense plan involves identifying vulnerable entry points in your home, such as doors and windows and securing them with locks, security bars or alarms. Establish a safe room in your home where you can retreat in case of a home invasion. This room should be equipped with a sturdy door, a phone to call for help and items that can be used for self-defense. Practice your home defense plan with all members of your household, ensuring everyone knows what to do in the event of an emergency.

Escape Routes: whether at home, work or in public, always be aware of the nearest exits and escape routes. In an emergency, knowing how to quickly leave the area can be the difference between safety and danger. Plan multiple escape routes for different scenarios, such as fires, natural disasters or active shooter situations. Regularly review and practice these escape plans so that they become second nature.

Communication Strategy: in an emergency, communication is key. Have a plan in place for contacting family members or emergency services. Ensure that everyone in your household knows how to use communication devices, such as phones or radios and establish a designated meeting point where you can regroup if separated. Additionally, consider carrying a whistle or personal alarm to signal for help if needed.

Legal Considerations: it's important to understand the legal implications of using self-defense techniques or tools. Laws regarding self-defense vary by location and using force inappropriately can have serious legal consequences. Educate yourself on the self-defense laws in your area, including what constitutes justifiable use of force and ensure that any defense tools you carry are legal to possess and use.

Concluding, mastering self-defense techniques involves a combination of physical training, strategic use of defense tools and thorough emergency planning. By being prepared and confident in your ability to protect yourself, you can navigate potentially dangerous situations with greater assurance and security. Self-defense is not just about physical prowess; it's about being proactive, aware and ready to act when necessary.

Securing your property and defending against potential threats is a multifaceted endeavor that requires careful planning, the right tools and ongoing vigilance. By understanding the principles of home security, constructing effective DIY security systems and developing personal self-defense skills, you can create a safe and resilient environment for yourself and your loved ones. As you implement these strategies, you'll gain not only the peace of mind that comes with a secure home but also the confidence to face any challenges that may arise.

Exercise Chapter 6
Designing a Home Security Plan

Objective: learn to design a comprehensive home security plan by assessing risks, identifying vulnerabilities and implementing layered security measures.

Materials Needed: notepad or digital document, pen or device for recording, tape measure, camera or smartphone (for documenting areas of concern) and a list of local crime statistics (optional).

1. Conduct a Risk Assessment: walk around your property and note all potential vulnerabilities. Consider factors like the visibility of valuable items, the accessibility of entry points (doors, windows, garage) and any patterns in your daily routine that could be exploited by intruders. Use your camera or smartphone to document areas of concern.

2. Evaluate Security Measures: examine existing security measures such as locks, fences, lighting and alarm systems. Identify any gaps or weaknesses. For example, check if windows and doors are adequately secured or if there are any dark areas around your property that could benefit from additional lighting.

3. Plan for Layered Security: based on your risk assessment, develop a layered security plan. This could include upgrading locks, installing surveillance cameras, adding motion-sensor lights or setting up a neighborhood watch. Prioritize actions based on the vulnerabilities you identified.

4. Implement Emergency Protocols: create an emergency response plan for various scenarios, such as a home invasion, fire or natural disaster. Determine escape routes, identify a safe room and establish communication strategies. Ensure that all household members are familiar with the plan and conduct regular drills.

5. Document and Review: write a brief report summarizing your home security plan. Include the steps you've taken, the rationale behind your decisions and any ongoing actions. Review this plan periodically, especially after any changes to your property or security needs.

Deliverable: submit a written report detailing your home security plan, including photos of identified vulnerabilities, a list of implemented security measures and a summary of your emergency protocols.

This exercise helps you create a proactive and customized home security plan, essential for protecting your property and ensuring the safety of your household.

Chapter 7
Emergency Communications

In times of crisis, effective communication is more than just a convenience, it's a lifeline. When natural disasters strike, infrastructure fails or you find yourself in a remote location, the ability to maintain communication can be the key to survival. Emergency communications encompass a range of tools and techniques designed to keep you connected when conventional systems are compromised or unavailable. These methods ensure that you can request help, stay informed and coordinate with others, no matter the situation.

Understanding the importance of emergency communication begins with recognizing its fundamental role in both personal safety and broader crisis management. Traditional communication networks, such as cellular or internet services, are often the first to fail during emergencies due to infrastructure damage or overloading. This vulnerability highlights the necessity of having reliable alternatives that can operate independently of these systems. Throughout this chapter, we will explore the tools and strategies that enable you to maintain vital connections in any scenario, emphasizing the importance of preparedness and the proper use of available resources.

Importance of Emergency Communications

In any emergency situation, communication becomes the lifeline that connects people to help, information and coordination efforts. The importance of emergency communications cannot be overstated, as it plays a critical role in ensuring safety, facilitating rescue operations and enabling informed decision-making. When traditional communication networks fail or become unreliable due to infrastructure damage or overloading, having alternative methods of communication is not just beneficial, it is essential.

The first and perhaps most significant aspect of emergency communication is its role in maintaining contact with emergency services, family members and other critical contacts. During crises like natural disasters, accidents or security threats, the ability to quickly reach out for help can be the difference between life and death. Emergency communication systems, such as two-way radios, satellite phones or even simple signaling devices, allow for the rapid dissemination of distress signals and the coordination of rescue operations. In these scenarios, communication tools must be reliable, accessible and easy to use under stress, ensuring that even those with minimal training can operate them effectively.

Furthermore, communication during emergencies is crucial for situational awareness. Access to real-time information allows individuals and organizations to understand the scope of the emergency, track the development of events and respond appropriately. Whether it's receiving weather alerts, evacuation orders or updates on the status of critical infrastructure, effective communication ensures that people are informed and can take the necessary actions to protect themselves and others. In situations where misinformation can lead to panic or poor decision-making, reliable communication channels help maintain order and provide clarity.

Another key aspect of emergency communications is its role in enabling coordination among response teams and between different groups of people. Whether you are coordinating with family members to ensure everyone is safe or working with a larger community to organize resources and efforts, communication is vital. For example, during a large-scale disaster, emergency response teams, including police, fire services and medical personnel, rely heavily on communication networks to share information, allocate resources and manage the response effectively. Without robust communication, these efforts can become disjointed, leading to inefficiencies and potentially endangering lives.

In addition to these immediate benefits, emergency communication systems also contribute to long-term resilience. By establishing and maintaining these systems, communities can better prepare for future emergencies, reducing the impact of such events. Regular drills and the integration of communication plans into broader emergency preparedness strategies ensure that when a crisis does occur, everyone knows their role and how to use the communication tools available to them.

Finally, it's important to recognize that emergency communications are not limited to technology. Traditional methods such as visual and acoustic signals, as well as predefined meeting points and communication plans, play a crucial role in ensuring that people can still communicate even when electronic devices are unavailable or inoperable. These methods provide a simple yet effective way to maintain contact and ensure that everyone involved in an emergency situation can stay connected.

In summary, the importance of emergency communications lies in its ability to save lives, provide situational awareness, enable coordinated responses and build long-term resilience. By understanding and implementing effective communication strategies, individuals and communities can significantly improve their preparedness for and response to emergencies, ensuring that they are ready to face the challenges that such situations present.

Radios and Transmitters

Radios and transmitters are fundamental tools in emergency communications, offering reliable means of staying connected when conventional communication systems fail. Whether you're dealing with a natural disaster, a widespread power outage or a remote expedition, understanding the different types of radios, how to build a basic transmitter and the best practices for usage and maintenance is crucial for ensuring effective communication in emergencies.

Types of Radios

The world of radios is vast, with various types designed to meet different communication needs. In an emergency, selecting the right type of radio can make all the difference. Here are the primary types of radios used in emergency communication:

- **Handheld Two-Way Radios (Walkie-Talkies)**: these are portable, battery-operated devices that allow two-way communication over short distances, typically a few miles. They are commonly used by families, small groups and emergency responders for quick and direct communication. Walkie-talkies operate on specific radio frequencies, often in the Family Radio Service (FRS) or General Mobile Radio Service (GMRS) bands.
- **Amateur (Ham) Radios**: ham radios are a versatile option for emergency communication, capable of long-distance communication across various frequencies. Licensed ham radio operators can use these devices to communicate locally, nationally or even internationally, depending on the equipment and atmospheric conditions. Ham radios can operate on VHF (Very High Frequency) and UHF (Ultra High Frequency) bands, as well as HF (High Frequency) bands, which are ideal for long-range communication.
- **CB Radios (Citizens Band)**: CB radios are another popular choice for emergency communication. They operate on 40 channels within the 27 MHz band and are commonly used by truckers, off-roaders and emergency services. CB radios have a range of about 3 to 20 miles, depending on terrain and atmospheric conditions and do not require a license to operate.
- **Marine Radios**: designed for communication on water, marine radios operate on the VHF band and are essential for anyone navigating rivers, lakes or oceans. These radios are crucial for contacting the Coast Guard, other vessels or emergency services during maritime emergencies.
- **Satellite Radios**: satellite radios offer global communication capabilities, making them ideal for use in remote areas where other forms of communication may be unreliable or unavailable. They connect to satellites orbiting the Earth, providing a stable link regardless of geographical location. Satellite radios are often used by adventurers, explorers and emergency responders working in remote or disaster-stricken areas.

Building a Transmitter

Building a basic radio transmitter can be an invaluable skill, especially in situations where traditional communication networks are down and you need to send a distress signal or communicate over long distances. While building a sophisticated transmitter requires technical expertise, a basic AM or FM transmitter can be constructed with relatively simple components. Here's a general overview of how to build a basic transmitter:

1. **Gather Materials**: to build a basic AM transmitter, you'll need an oscillator (which can be made using a simple transistor circuit), a microphone (for voice input), a power source, an antenna and some connecting wires. For FM transmitters, the setup is similar, but the circuit design will differ slightly to modulate the frequency.
2. **Construct the Oscillator Circuit**: the oscillator is the heart of your transmitter, generating the carrier signal that will be modulated with your voice or data. You can build a basic oscillator using a transistor, resistors, capacitors and inductors arranged in a specific circuit configuration. This circuit will oscillate at a specific frequency, which determines your transmission frequency.
3. **Integrate the Microphone**: the microphone is used to capture your voice, converting sound waves into an electrical signal. This signal is then fed into the oscillator circuit, where it modulates the carrier wave. The modulation process combines the audio signal with the carrier wave, allowing it to be transmitted over the air.
4. **Connect the Antenna**: the antenna is critical for broadcasting your signal. It should be a length of wire or a telescoping antenna, tuned to the wavelength of your carrier frequency. The antenna radiates the modulated signal into the air, where it can be picked up by a receiver tuned to the same frequency.
5. **Power the Circuit**: finally, connect your circuit to a power source, such as a battery. Ensure that all connections are secure and that the components are properly aligned. Once powered, your transmitter will begin broadcasting the modulated signal, which can be received by any compatible radio within range.
6. **Test and Adjust**: testing your transmitter is crucial to ensure it works correctly. Tune a radio to the frequency of your transmitter and listen for your broadcast. If the signal is weak or unclear, you may need to adjust the oscillator circuit or the length of the antenna.

Usage and Maintenance

Using radios and transmitters effectively during emergencies requires not only familiarity with the equipment but also regular maintenance to ensure they are operational when needed. Here are some best practices for usage and maintenance:

- **Regular Testing**: periodically test all communication equipment to ensure it is functioning correctly. This includes checking the battery levels, ensuring the antenna is properly connected and verifying that you can transmit and receive signals clearly. Regular tests help identify issues before they become critical during an emergency.

- **Battery Management**: keep spare batteries on hand and regularly check rechargeable batteries to ensure they hold a charge. Consider using solar chargers or hand-crank generators to maintain power to your devices during prolonged outages.
- **Antenna Care**: the antenna is crucial for effective transmission and reception. Regularly inspect it for damage and ensure it is properly positioned and connected. In the case of portable radios, extend the antenna fully during use to maximize range.
- **Stay Informed**: familiarize yourself with the operating frequencies and protocols of local emergency services, ham radio networks and other relevant communication channels. Knowing which frequencies to monitor or transmit on can be vital during an emergency.
- **Emergency Drills**: incorporate radio use into your emergency drills. Practice using your equipment under simulated emergency conditions to ensure that all users are comfortable and proficient in its operation.
- **Documentation**: keep a logbook of your radio and transmitter settings, frequencies and maintenance activities. This documentation can be invaluable for troubleshooting issues and ensuring consistent communication practices.

In conclusion, radios and transmitters are essential tools in any emergency communication strategy. By understanding the different types of radios, learning how to build a basic transmitter and maintaining your equipment properly, you can ensure reliable communication during emergencies. This not only enhances your personal safety but also strengthens your ability to assist others and coordinate with emergency services in times of crisis.

Visual and Acoustic Signaling

In emergency situations, when conventional communication methods are unavailable or impractical, visual and acoustic signaling can be vital. These techniques allow individuals to convey messages over distances, ensuring that communication remains possible even in challenging conditions. Understanding the principles behind these signaling methods, the tools you can use and how to ensure safety while communicating are essential components of a comprehensive emergency preparedness plan.

Signaling Techniques

Visual and acoustic signaling methods are designed to transmit messages without relying on electronic devices. These techniques are especially useful when silence, simplicity or non-verbal communication is required.

Visual Signaling:
- **Flag Signals**: traditionally used in maritime and military contexts, flag signaling can be adapted for emergency use. Different flag positions can convey various messages, with the most basic being the universal distress signal: waving a flag or any bright cloth.
- **Light Signals**: reflective surfaces like mirrors or the use of flashlights can effectively communicate over long distances, especially at night. The key is to use patterns such as Morse code, where short and long

flashes represent dots and dashes, respectively. The universally recognized SOS signal (••• --- •••) is an essential skill, providing a way to signal distress.
- **Hand Signals**: silent and efficient, hand signals are critical in scenarios requiring stealth or when verbal communication is impossible. These signals need to be pre-agreed within your group to avoid confusion, with simple gestures indicating directions, warnings or actions.
- **Smoke Signals**: using smoke to create visual signals has been a traditional method among various cultures. The key to effective smoke signaling is control, by covering and uncovering the fire to produce distinct puffs, you can send messages over long distances. The classic emergency signal is three puffs of smoke in quick succession.
- **Ground-to-Air Signals**: when stranded or lost, especially in open areas, ground-to-air signals can be your best bet for being spotted by search and rescue teams. These large, contrasting symbols, created using materials like rocks, branches or even drawn in sand, can convey messages such as "need help" or "safe."

Acoustic Signaling:
- **Whistles**: a whistle is a simple yet powerful tool for signaling. The sound can carry over long distances, cutting through noise and environmental conditions. The international distress signal for a whistle is three short blasts.
- **Voice Commands**: though limited in range and affected by environmental factors, shouting or specific voice commands can be effective for short-range communication. However, it's best to use this method sparingly to avoid strain and ensure clarity.
- **Horn or Bell Signals**: these tools are commonly found in vehicles and boats, with specific patterns of sounds used to convey warnings or attract attention. In an emergency, repeating the horn or bell sound in sets of three can indicate distress.
- **Drum Signals**: historically used for communication across distances in various cultures, drum signals can be adapted for modern emergencies, especially in dense environments like forests where other signals might not be effective.
- **Gunshots**: while generally a last resort due to legal and safety concerns, gunshots can be used to signal distress in remote areas. The signal typically involves three shots fired in quick succession.

Signaling Tools

To enhance the effectiveness of your visual and acoustic signals, several tools are available that can increase range, visibility and audibility:

- **Mirrors**: a small, compact mirror can be a lifesaving tool, capable of reflecting sunlight over vast distances. Signal mirrors are specifically designed for this purpose and should be part of any emergency kit.
- **Flares**: whether handheld or launched, flares are highly visible in low-light conditions and can signal distress from far away. Their bright, intense light makes them ideal for nighttime emergencies or in foggy conditions.

- **Glow Sticks**: these are practical for nighttime signaling. Their steady light is less intense than flares but can still be used effectively to mark locations or signal to nearby searchers.
- **Signal Whistles**: designed to emit a loud, piercing sound, signal whistles are essential for emergency situations. Their compact size makes them easy to carry, ensuring they're accessible when needed.
- **Air Horns**: powerful and attention-grabbing, air horns are commonly used in maritime and industrial settings. They can be included in your emergency kit to signal distress or alert others in noisy environments.
- **Flag Kits**: a flag kit, complete with brightly colored flags and poles, allows for easy visual signaling in open spaces or on water. These kits are particularly useful in maritime contexts.

Communication Safety

When using visual and acoustic signaling, it's crucial to ensure that your communication is both clear and safe. Miscommunication or using the wrong signal can lead to serious consequences.

Signal Clarity: practice using your signaling tools and techniques to ensure they are clear and easily understood. In stressful situations, familiarity with these methods is vital to avoid mistakes.

Avoiding Detection by Hostiles: in certain situations, such as during civil unrest or in hostile territories, signaling might attract unwanted attention. Use low-profile tools and discreet signals when necessary to avoid revealing your position.

Environmental Considerations: be mindful of the environment when signaling. Weather conditions like fog, rain or wind can affect the visibility and audibility of your signals. Adapt your techniques accordingly.

Legal and Ethical Considerations: some signaling methods, like gunshots or flares, should only be used in genuine emergencies due to their potential legal implications and the risks they pose.

Mastering visual and acoustic signaling techniques is a crucial aspect of emergency preparedness. These methods offer reliable ways to communicate when other systems fail, enhancing your ability to stay connected and safe during crises.

In conclusion, mastering emergency communications requires a multi-faceted approach that includes understanding the importance of communication, selecting the right tools and being proficient in various signaling techniques. By combining radios, transmitters and visual or acoustic signals, you can ensure that you are prepared for any situation where reliable communication is critical. This preparedness not only enhances your safety but also contributes to the effectiveness of broader emergency response efforts. Whether you are planning for a natural disaster, a remote expedition or any scenario where conventional communication might fail, the knowledge and skills covered in this chapter will equip you to stay connected and informed, no matter the circumstances.

Exercise Chapter 7
Building and Testing a Basic Radio Transmitter

Objective: learn how to build and test a basic radio transmitter for emergency communications.

Materials Needed: transistor (NPN type), resistors (1 kΩ and 10 kΩ), capacitor (100 pF), microphone, battery (9V), antenna (wire), breadboard, connecting wires and a multimeter.

1. Assemble the Oscillator Circuit: start by placing the transistor on the breadboard. Connect the 1 kΩ resistor between the collector (C) of the transistor and the positive terminal of the battery. Next, connect the 10 kΩ resistor between the base (B) of the transistor and the positive terminal. Finally, place the capacitor between the base and the emitter (E) and connect the emitter directly to the negative terminal of the battery.

2. Connect the Microphone: attach the microphone to the circuit by connecting one end to the base of the transistor and the other to the ground (negative terminal of the battery). This setup will allow the microphone to modulate the signal generated by the oscillator circuit.

3. Attach the Antenna: connect the antenna to the collector of the transistor. The antenna should be a wire of appropriate length (typically 1 meter or more) to effectively broadcast the signal.

4. Power the Circuit: connect the battery to the circuit to power the transmitter. Ensure that all connections are secure and that the components are properly aligned.

5. Test the Transmitter: use a nearby AM or FM radio tuned to an open frequency. Turn on the transmitter and speak into the microphone. Adjust the radio until you hear the broadcasted signal clearly. Use the multimeter to check for any voltage drops or irregularities in the circuit.

6. Troubleshoot and Optimize: if the signal is weak or distorted, try adjusting the antenna length or repositioning the components on the breadboard. Check all connections for stability and re-test the circuit until you achieve a clear signal.

Deliverable: write a short report detailing your experience building the radio transmitter. Include any challenges you encountered, how you resolved them and photos of your setup.

This exercise provides hands-on experience in constructing a basic radio transmitter, a crucial skill for emergency communication preparedness.

Chapter 8
Building Emergency Shelters

In any emergency, having a reliable shelter is paramount to ensuring your safety and survival. Whether faced with natural disasters, such as hurricanes and earthquakes or human-made crises, such as civil unrest, the ability to construct or find an emergency shelter can make the difference between life and death. Emergency shelters provide protection from harsh weather, security from potential threats and a secure space to rest and recover. They range from simple temporary structures to more complex underground bunkers, each designed to meet different needs and circumstances.

The importance of emergency shelters cannot be overstated. When disaster strikes, your home may no longer be safe and having a pre-built or quickly assembled shelter can save lives. Emergency shelters vary widely, from simple makeshift structures that can be erected in minutes to more permanent underground bunkers designed to withstand severe conditions. The type of shelter you choose to build or utilize will depend on various factors, including the nature of the threat, the environment you are in and the resources at your disposal.

Introduction to Emergency Shelters

In the face of an emergency, whether it be natural disasters, conflicts or other unexpected crises, the ability to secure adequate shelter is fundamental to survival. Emergency shelters provide a safe haven, protecting individuals from the elements, environmental hazards and potential threats. The concept of an emergency shelter goes beyond just a place to sleep, it is a vital component of survival strategy, offering security, psychological comfort and a base from which to assess and respond to ongoing challenges.

Understanding the importance of emergency shelters begins with recognizing the diverse types of emergencies that may necessitate their use. Natural disasters such as hurricanes, tornadoes, earthquakes and floods can render conventional housing unsafe or uninhabitable within moments. In these situations, emergency shelters, whether improvised or pre-constructed, become essential for survival. Similarly, in scenarios involving civil unrest, pandemics or large-scale power outages, having a secure and well-supplied shelter can provide the necessary protection against physical threats and health risks.

Emergency shelters can be categorized into two main types: temporary and permanent. Temporary shelters, such as tents or makeshift structures, are designed for short-term use and can be quickly assembled in the immediate aftermath of a disaster. These shelters are lightweight, portable and easy to set up, making them ideal for situations where mobility and speed are critical. On the other hand, permanent shelters, such as underground bunkers or reinforced rooms, are designed to offer long-term protection and can withstand more severe conditions. These shelters provide a more secure and comfortable environment for extended stays but require significant planning and resources to construct.

Selecting the right site for an emergency shelter is a crucial aspect of planning and construction. The location of a shelter can significantly impact its effectiveness and the safety of its occupants. Factors such as proximity to water sources, the risk of flooding, exposure to wind or falling debris and accessibility to escape routes must all be carefully considered. In addition, the shelter should be located in an area that minimizes visibility to potential threats while ensuring that it remains accessible and functional for the duration of the emergency.

The benefits of having a well-constructed and strategically located emergency shelter are manifold. Beyond providing physical protection, a shelter offers psychological security, which is crucial during times of crisis. The sense of safety that a reliable shelter provides can help maintain morale, reduce stress and allow individuals to focus on critical tasks such as sourcing food, water and medical supplies. Furthermore, shelters serve as a base of operations, enabling individuals or groups to plan their next steps, communicate with others and manage resources effectively.

In emergency preparedness, the ability to build or secure an appropriate shelter is as important as having access to food, water and medical supplies. This chapter will guide you through the different types of emergency shelters, from the most basic temporary setups to more complex underground bunkers. We will explore the practical considerations involved in site selection, the construction process and the ongoing maintenance required to ensure that your shelter remains safe and functional. By mastering these skills, you will be better equipped to protect yourself and your loved ones in any emergency, ensuring that you have a secure and resilient refuge in times of need.

Building an Underground Shelter

Constructing an underground shelter is a significant undertaking, but it offers unparalleled protection and security in various emergency situations. Whether designed to protect against natural disasters, nuclear fallout or other catastrophic events, an underground shelter provides a safe, concealed space that can sustain life for extended periods. This section will explore the critical aspects of building an underground shelter, including the design considerations, excavation and construction processes, as well as essential elements like waterproofing and security.

Shelter Design

The design of an underground shelter is the foundation of its effectiveness. A well-thought-out design considers not only the size and layout but also factors such as ventilation, accessibility and comfort. The first step in designing an underground shelter is determining its primary purpose. For instance, a shelter intended for short-term use during a tornado will have different design requirements than one meant for long-term habitation in the aftermath of a nuclear event.

Size and Layout: the size of your shelter will depend on the number of occupants, the duration of stay and the supplies needed. A basic shelter should include areas for sleeping, storage and essential activities such as cooking and sanitation. The layout should maximize space efficiency while ensuring that occupants can move comfortably. It's also crucial to consider future expansion, especially if the shelter is intended for long-term use.

Ventilation and Air Filtration: proper ventilation is vital for maintaining a livable environment inside an underground shelter. Fresh air intake and exhaust systems must be included in the design to prevent the buildup of carbon dioxide and other harmful gases. In more advanced shelters, air filtration systems can be installed to remove contaminants from the air, providing protection against chemical, biological or radioactive threats.

Structural Integrity: the shelter must be designed to withstand the pressures exerted by the surrounding earth as well as potential impacts from above. Reinforced concrete is commonly used for its strength and durability. The walls, roof and floor must all be constructed to resist collapse and protect against external forces, whether from natural events like earthquakes or human-made threats such as explosions.

Accessibility: the entrance to the shelter should be well-concealed yet easily accessible in an emergency. Consider designing multiple entry points if possible, including an emergency exit to ensure that occupants are not trapped inside. The entrance should also be protected by a blast door or similar barrier to prevent unauthorized access and shield against explosions.

Excavation and Construction

Once the design is finalized, the next phase is excavation and construction. Building an underground shelter requires significant planning and resources, especially if the shelter is to be large or deeply buried.

Excavation: the first step in construction is excavating the site. This involves removing the soil and rock to create a cavity large enough to accommodate the shelter. The depth of the shelter will depend on its intended purpose; deeper shelters offer more protection but are more challenging and expensive to build. It's essential to consider soil stability and the presence of groundwater during excavation. Hiring professionals with experience in excavation and underground construction is highly recommended to avoid structural issues.

Foundation and Structure: after excavation, a strong foundation is laid to support the shelter. This is typically constructed from reinforced concrete, which provides the necessary strength and durability. The walls and roof are then built, with attention to ensuring they can withstand external pressures and potential impacts. The structure must be carefully sealed to prevent water infiltration and ensure long-term stability.

Utilities and Systems: during construction, essential systems such as electrical wiring, plumbing and ventilation must be installed. These systems need to be durable and easy to maintain, as access for repairs may be limited. Backup power sources, such as generators or solar panels, should be considered to ensure that the shelter remains functional even if the primary power supply is disrupted.

Waterproofing and Security

Waterproofing and security are critical components of any underground shelter, ensuring that it remains dry and safe from external threats.

Waterproofing: one of the most significant challenges in building an underground shelter is preventing water infiltration. The shelter must be thoroughly waterproofed to avoid flooding and dampness, which can lead to mold, structural damage and the degradation of stored supplies. Waterproofing involves applying specialized coatings or membranes to the exterior walls and roof, as well as installing drainage systems to divert water away from the shelter. It's essential to test the waterproofing thoroughly before the shelter is occupied.

Security Measures: security is a key concern, particularly for shelters intended to protect against human threats. The shelter should be equipped with robust doors that can resist forced entry, as well as security systems like cameras, alarms and motion detectors. Concealing the shelter's location is also an effective security measure; it should blend into its surroundings or be hidden by natural features. Inside the shelter, provisions for self-defense, such as reinforced doors and secure storage for supplies, should be considered.

Environmental Control: maintaining a stable internal environment is crucial for the comfort and safety of the occupants. This includes controlling temperature, humidity and air quality. Insulation should be incorporated into the shelter's design to regulate temperature, while dehumidifiers can help control moisture levels. Advanced shelters may also include systems for monitoring and adjusting air quality, ensuring a constant supply of clean air.

Building an underground shelter is a complex but rewarding project that can provide unparalleled protection and peace of mind in an emergency. By carefully planning the design, executing the construction with precision and incorporating robust waterproofing and security measures, you can create a shelter that offers safety and comfort in even the most extreme conditions. Whether as a precautionary measure or a critical part of your emergency preparedness plan, an underground shelter is an investment in safety and resilience.

Temporary and Portable Shelters

Temporary and portable shelters are essential tools for survival in various emergency situations. Whether you're facing a natural disaster, forced to evacuate your home or venturing into the wilderness, these shelters provide immediate protection against the elements and other potential hazards. Their versatility and ease of setup make them invaluable assets in any emergency preparedness plan. This section explores the different types of temporary and portable shelters, the materials required for their construction and the techniques for effective setup and takedown.

Tents and Makeshift Shelters

Temporary shelters can range from commercial tents designed for quick deployment to improvised structures made from materials at hand. The choice of shelter depends on the environment, the duration of stay and the available resources.

Tents: tents are one of the most common forms of temporary shelters due to their portability, ease of setup and ability to provide reliable protection against the elements. Modern tents come in various shapes and sizes, from small, lightweight backpacking tents to larger family tents that offer more space and comfort. Most tents are designed with water-resistant or waterproof materials, ensuring that they remain dry inside during rain. Tents also typically feature ventilation systems, such as mesh windows or vents, to reduce condensation and maintain airflow.

Dome Tents: dome tents are popular for their stability and resistance to wind. Their curved design allows rain and snow to slide off easily, preventing accumulation that could cause collapse. They are easy to set up, often requiring just a few poles and stakes.

Tunnel Tents: tunnel tents are elongated and provide more interior space, making them ideal for larger groups or extended stays. They are designed to be aerodynamically efficient, offering good resistance to wind when properly oriented.

Pop-Up Tents: these are ultra-portable and can be set up in seconds. Pop-up tents are ideal for short-term use, such as during a brief camping trip or when quick shelter is needed. However, they may not be as durable or weather-resistant as other types of tents.

Makeshift Shelters: in situations where tents are unavailable, makeshift shelters constructed from natural or scavenged materials can provide critical protection. These shelters are typically more rudimentary but can be highly effective when built correctly.

Lean-To: a lean-to is a simple structure made by leaning branches, poles or other materials against a support, such as a tree or a ridgepole. The roof is then covered with leaves, tarps or other waterproof materials to create a slanted surface that deflects rain and wind. Lean-tos are quick to construct and can be expanded or reinforced as needed.

Debris Hut: a debris hut is a survival shelter built by creating a frame of sticks or branches and covering it with leaves, grass and other insulating materials. The debris provides protection from the wind and cold, making it ideal for colder climates. The entrance should be small to minimize heat loss and the shelter should be densely packed with debris for maximum insulation.

A-Frame Shelter: similar to a lean-to, an A-frame shelter uses a ridgepole supported at both ends, with branches or poles forming the sides. The frame is then covered with insulating materials. This design provides better protection from the elements than a lean-to, especially in snowy or windy conditions.

Required Materials

The materials needed for temporary and portable shelters vary depending on the type of shelter and the environment. Here are some common materials used in shelter construction:

For Tents:
- **Fabric**: durable, weather-resistant fabrics such as nylon or polyester are used for the tent body and rainfly. These materials are lightweight yet strong, providing protection from rain, wind and UV rays.
- **Poles**: tent poles are typically made from aluminum, fiberglass or carbon fiber. Aluminum poles are the most common, offering a balance of strength, weight and flexibility.
- **Stakes and Guy Lines**: stakes anchor the tent to the ground, while guy lines provide additional stability. Stakes can be made from metal, plastic or wood, depending on the ground conditions.
- **Groundsheet**: a groundsheet or footprint protects the tent floor from abrasion and moisture, extending the tent's lifespan and enhancing comfort.

For Makeshift Shelters:
- **Natural Materials**: branches, leaves, grass and stones are often used in the construction of makeshift shelters. These materials are readily available in most environments and can provide adequate protection with proper assembly.
- **Tarps and Ropes**: tarps are invaluable in makeshift shelter construction, providing waterproofing and wind resistance. Ropes or paracord are essential for securing materials and creating a sturdy framework.
- **Tools**: basic tools like a knife, saw or hatchet can greatly facilitate the construction of makeshift shelters, allowing you to cut and shape materials more efficiently.

Setup and Takedown Techniques

Setting up and taking down temporary and portable shelters quickly and efficiently is crucial, especially in emergencies where time and resources may be limited. Proper technique ensures that the shelter is secure, weatherproof and ready to provide protection.

Site Selection: choose a location that offers natural protection from the elements, such as a site with good drainage, some windbreak and natural cover. Avoid areas prone to flooding, falling branches or other hazards. The ground should be as level as possible to ensure stability and comfort.

Tent Setup:
1. **Unpack and Lay Out the Components**: identify and organize all parts of the tent, including poles, stakes, rainfly and the tent body.
2. **Assemble the Poles**: connect the poles and insert them into the designated sleeves or clips on the tent body. Raise the tent by securing the poles in the grommets or corner sockets.
3. **Stake the Tent**: secure the tent to the ground using stakes, driving them into the soil at a 45-degree angle for maximum hold. Use the guy lines to tension the tent and stabilize it against wind.
4. **Add the Rainfly**: attach the rainfly over the tent, ensuring it is taut and covers all areas exposed to potential rainfall. Secure the fly with additional stakes or lines as needed.

Makeshift Shelter Setup:
1. **Construct the Frame**: use branches, poles or other sturdy materials to create the framework of the shelter. For a lean-to, this might involve leaning branches against a tree or ridgepole. For an A-frame, position the ridgepole between two supports.
2. **Add Insulation and Covering**: cover the frame with leaves, grass or tarps to provide insulation and waterproofing. Ensure that the covering is dense enough to block wind and rain.
3. **Reinforce and Secure**: use ropes or vines to secure the covering to the frame. If possible, anchor the shelter to the ground using stakes or weighted objects.

Takedown and Packing:
1. **Disassemble Carefully**: start by removing stakes, guy lines and the rainfly. Take care to fold and pack each component neatly to avoid damage and ensure easy setup next time.
2. **Inspect for Damage**: before packing away your shelter, inspect all components for wear or damage. Repair or replace any compromised parts to ensure the shelter remains reliable.
3. **Leave No Trace**: especially when using natural sites, ensure that you leave the area as you found it. Pack out all non-biodegradable materials and restore the site as much as possible to minimize your environmental impact.

Temporary and portable shelters are indispensable in emergency situations, providing immediate, adaptable protection when it's needed most. Whether you're setting up a tent in a safe location or improvising a makeshift shelter from natural materials, knowing how to efficiently construct, maintain and disassemble these shelters can make all the difference in an emergency. These shelters are not only practical but also empower you with the skills and confidence to face adverse conditions with resilience and resourcefulness.

In conclusion, mastering the art of building emergency shelters is a critical component of survival preparedness. Whether constructing a permanent underground bunker or quickly assembling a temporary shelter, the ability to protect yourself and your loved ones from the elements and potential threats is invaluable. By understanding the different types of shelters, carefully planning their construction or setup and being prepared with the necessary materials and skills, you can ensure that you are ready for any emergency situation that may arise.

Exercise Chapter 8
Constructing and Testing a Lean-To Shelter

Objective: learn how to build and test a lean-to shelter using natural materials, focusing on structural stability, weatherproofing and comfort.

Materials Needed: long, sturdy branches (for the main frame), smaller branches (for the roof structure), leaves, grass or pine needles (for insulation), rope or vines (optional, for securing materials) and a groundsheet or tarp.

1. Assemble the Frame: start by selecting two strong, vertical support branches or trees. Lay a long horizontal branch (the ridgepole) across the supports at a slight angle to create a sloped roof. Ensure that the ridgepole is securely tied or wedged to prevent movement.

2. Construct the Roof: lean smaller branches against the ridgepole at regular intervals, creating a slanted roof structure. These branches should be close together to provide adequate support for the covering material.

3. Add Insulation: cover the roof structure with a thick layer of leaves, grass or pine needles. The insulation should be at least a foot thick to provide protection from rain and wind. Pack the materials tightly to minimize gaps and enhance water resistance.

4. Secure the Shelter: use rope or vines to tie the covering material to the structure, ensuring that it stays in place even in strong winds. If available, place a groundsheet or tarp on the floor of the shelter to provide added insulation from the cold ground.

5. Test the Shelter: after completing the shelter, check its stability by gently pushing it. Simulate rain to test for leaks and make necessary adjustments for better stability and waterproofing.

6. Troubleshoot and Optimize: if the shelter is unstable or leaks, reinforce it with more branches or secure loose materials. Adjust the roof slope if needed to improve water runoff.

Deliverable: write a brief report on your experience building the lean-to shelter. Highlight any challenges, how you resolved them and include photos of the finished shelter. Reflect on how well the shelter provided protection and comfort.

This exercise provides hands-on experience in constructing a lean-to shelter, an essential skill for survival in wilderness and emergency situations.

Chapter 9
Wilderness Survival Techniques

Surviving in the wilderness requires not just physical endurance but also a deep understanding of the natural environment and the skills to navigate it safely. Whether you are an adventurer exploring remote areas or find yourself unexpectedly stranded in the wild, knowing how to survive can mean the difference between life and death. Wilderness survival encompasses a wide range of skills, from finding and purifying water to navigating without a map. However, some of the most critical aspects include building shelter, starting a fire and effectively using available resources. This chapter will explore these essential survival techniques, focusing on their practical application in real-world scenarios.

Survival in the wilderness is not just about reacting to the situation at hand but also about preparation and knowledge. By understanding the importance of survival skills, equipping yourself with the right tools and planning effectively, you can significantly increase your chances of staying safe and eventually finding your way back to civilization. This chapter will guide you through the basics of wilderness survival, offering practical advice and techniques that you can apply in various situations.

Introduction to Wilderness Survival

Surviving in the wilderness is an ancient skill set that has been essential to human survival for thousands of years. In today's world, while many of us are far removed from the need to fend for ourselves in nature, the ability to survive in the wilderness remains a crucial skill for anyone who ventures into the great outdoors.

Whether you are a seasoned adventurer, a weekend hiker or someone who finds themselves in an unexpected situation, understanding the basics of wilderness survival can make the difference between life and death.

The importance of wilderness survival skills cannot be overstated. In an era where technology and modern conveniences dominate our daily lives, it is easy to forget that nature is unpredictable and, at times, unforgiving. Natural disasters, accidents or simply getting lost can quickly turn a pleasant outing into a life-threatening situation. In these moments, your knowledge of survival techniques – such as finding shelter, sourcing water and creating fire – becomes your most valuable asset. These skills are not just about enduring the wilderness; they are about thriving and maintaining a level of comfort and safety until help arrives or you can navigate back to civilization.

Importance of Survival Skills

Survival skills are essential for anyone who spends time in nature, whether for recreation or work. These skills are not just practical; they are also empowering. Knowing that you can rely on yourself to find food, water and shelter and to stay safe in the wilderness, builds confidence and reduces fear in challenging situations. Moreover, survival skills teach resilience and adaptability, qualities that are not only crucial in the wild but also in everyday life. They enable you to stay calm and focused in the face of adversity, making logical decisions based on the resources available to you.

Additionally, mental preparedness plays a critical role in wilderness survival. Being mentally prepared means staying calm, focused and making rational decisions even when faced with unexpected challenges. Understanding that survival situations can be mentally taxing and preparing yourself to stay calm and focused can greatly improve your ability to make sound decisions under pressure. Regularly practicing survival skills, even in a controlled environment, can build your confidence and ensure that you are ready to act effectively in a real emergency.

In a broader sense, wilderness survival skills also foster a deeper connection with nature. When you know how to read the land, understand animal behavior and utilize natural resources, you develop a greater appreciation for the environment and its ecosystems. This knowledge can inspire more responsible and sustainable practices, both in the wild and in daily life.

Basic Tools

Equipping yourself with the right tools is a fundamental part of wilderness survival. While the human body and mind are incredibly resourceful, having certain tools can dramatically increase your chances of survival. The most basic survival tools include a reliable knife, fire-starting equipment, a compass and a first-aid kit.

Knife: a knife is arguably the most important tool in a survival kit. It can be used for a multitude of tasks, from cutting rope and preparing food to carving wood and creating makeshift tools. A good survival knife should be durable, easy to sharpen and capable of handling heavy use.

Fire-Starting Equipment: fire is crucial for warmth, cooking, water purification and signaling for help. Carrying reliable fire-starting tools, such as waterproof matches, a lighter or a ferrocerium rod, is essential. These tools should be stored in a waterproof container to ensure they remain functional in wet conditions.

Compass and Map: navigation is key to finding your way out of the wilderness or to locating resources such as water or shelter. A compass and a topographic map of the area you are exploring are invaluable tools for staying oriented and avoiding getting lost. While GPS devices are useful, they should not be solely relied upon, as they can fail due to battery depletion or loss of signal.

First-Aid Kit: injuries are common in the wilderness and a well-stocked first-aid kit can prevent minor injuries from becoming serious. Your kit should include bandages, antiseptics, pain relievers and any personal medications you might need. Knowing basic first-aid procedures, such as how to treat cuts, sprains and insect bites, is equally important.

Planning and Preparation

Preparation is the cornerstone of survival. Before venturing into the wilderness, thorough planning can significantly reduce the risks associated with outdoor activities. This includes understanding the environment you will be entering, such as the climate, terrain and potential hazards. Preparing for the unexpected – such as sudden weather changes, wildlife encounters or navigation challenges – can be the key to your survival.

One of the most critical aspects of preparation is informing someone of your plans. Before you set out, always let a trusted person know where you are going, your intended route and your expected return time. This ensures that if something goes wrong, search and rescue teams have a starting point for their efforts.

In addition to sharing your plans, packing the right gear is essential. This includes not only your basic survival tools but also appropriate clothing, food and water. Layered clothing made from moisture-wicking and insulating materials can protect you from hypothermia or overheating. Packing high-energy, non-perishable food items and a means of water purification, such as a filter or purification tablets, is crucial for sustaining yourself over an extended period.

Finally, mental preparation is just as important as physical preparation. Understanding that survival situations can be mentally taxing and preparing yourself to stay calm and focused can greatly improve your ability to make sound decisions under pressure. Regularly practicing survival skills, even in a controlled environment, can build your confidence and ensure that you are ready to act effectively in a real emergency.

Concluding, wilderness survival is about more than just enduring the elements; it's about being prepared, resourceful and resilient. By understanding the importance of survival skills, equipping yourself with the necessary tools and thoroughly planning and preparing for your journey, you can significantly enhance your chances of surviving and thriving in the wild.

Fire Starting

The ability to start a fire is one of the most crucial survival skills you can possess when venturing into the wilderness. Fire provides warmth, cooks food, purifies water and serves as a signal for rescue. In many survival situations, the success of your fire-making efforts can mean the difference between life and death. This section will delve into various fire-starting methods, the materials you need and important fire safety considerations to ensure you can confidently build and maintain a fire in any situation.

Traditional and Modern Methods

Fire starting techniques can be broadly categorized into traditional and modern methods, each with its own advantages and challenges. Understanding both allows you to adapt to the tools and materials available in your environment.

Friction-Based Methods: one of the oldest methods of fire-starting, friction-based techniques involve rubbing two materials together to create heat, which eventually ignites tinder. The bow drill is a common example of this method. It consists of a spindle (a straight stick) that rotates against a fireboard using a bow (a bent stick with a cord) to create friction. While effective, this method requires skill, patience and the right materials, making it less reliable for those unfamiliar with the technique.

Flint and Steel: another traditional method involves striking a piece of steel against a flint rock. The sparks generated by the impact can ignite a small pile of tinder. Flint and steel kits are compact, durable and reliable, even in wet conditions. However, mastering this method also requires practice to efficiently produce a flame.

Fire Plough: this is another ancient technique, where a stick (the plough) is rapidly rubbed against a groove in a softer piece of wood. The friction generates heat, which eventually produces embers that can ignite a tinder bundle. The fire plough is simple but physically demanding, requiring considerable effort and the right type of wood.

Matches: modern matches are a convenient and reliable fire-starting tool. They come in various types, including waterproof and windproof varieties designed for survival situations. Matches are easy to use and effective, but they have a limited supply and they can become useless if they get wet (unless they are waterproof).

Lighters: butane lighters are another modern fire-starting tool that is highly effective and easy to use. A lighter can produce hundreds of flames on a single fuel charge and works well in most conditions. However, lighters can fail in extremely cold temperatures or if they become wet.

Ferrocerium Rods (Ferro Rods): a ferro rod is a modern fire-starting tool made from a blend of metals that produces sparks when scraped with a metal striker. These rods are reliable, even in wet conditions and can generate thousands of sparks before wearing out. Ferro rods are lightweight, compact and a favorite among survivalists because they work well with various tinder materials.

Fire Starting Materials

Successfully starting a fire requires more than just knowing the methods, it also depends on using the right materials. Understanding the different types of materials involved in fire-starting is essential for ensuring that your fire ignites quickly and burns steadily.

Tinder: tinder is the material that catches the initial spark and starts the fire. It must be dry, finely shredded and highly flammable. Good natural tinder includes dry grass, leaves, bark and pine needles. You can also prepare tinder by shredding materials like cotton balls, lint or paper. In a survival situation, having pre-made tinder, such as char cloth or commercially available fire-starting cubes, can be invaluable.

Kindling: once the tinder is burning, kindling is used to build the fire. Kindling is made up of small sticks, twigs and branches that catch fire easily from the tinder and produce enough heat to ignite larger pieces of wood. The kindling should be dry and arranged in a way that allows air to flow through the fire, such as in a teepee or log cabin structure.

Fuel Wood: the final component of a sustainable fire is fuel wood, which consists of larger logs or branches. Fuel wood provides the long-lasting heat needed for cooking, warmth and maintaining the fire over extended periods. It's important to select dry, seasoned wood, as wet or green wood will produce more smoke and be harder to keep burning.

Fire Accelerants: in challenging conditions, fire accelerants can help get a fire going more quickly. These include substances like petroleum jelly (often applied to cotton balls), hand sanitizer or chemical fire starters. These materials are easy to ignite and can help sustain the flame long enough for the tinder and kindling to catch.

Fire Safety

While fire is an essential survival tool, it can also be dangerous if not handled properly. Understanding fire safety is crucial to prevent accidents and wildfires, especially in dry or windy conditions where a small spark can quickly escalate into a large fire.

Site Selection: choosing the right location for your fire is the first step in ensuring safety. The site should be on bare earth or sand, away from flammable materials like dry grass, leaves or overhanging branches. In some cases, it may be necessary to clear an area around the fire pit to create a buffer zone. Building your fire in an established fire ring, if available, is always a good practice.

Wind Protection: wind can spread flames and embers, so it's important to shield your fire from gusts. Building your fire against a natural windbreak, such as a rock or a hill or constructing a makeshift barrier can help keep the fire contained.

Fire Management: never leave a fire unattended, as it can quickly grow out of control. Keep your fire small and manageable, especially in dry conditions. Always have a method for extinguishing the fire close at hand, such as a bucket of water or a shovel for covering the fire with dirt.

Extinguishing the Fire: when you are finished with your fire, make sure it is fully extinguished. This involves pouring water over the fire, stirring the ashes and then adding more water until the ashes are cool to the touch. In remote or wilderness areas, this is crucial to prevent unintended forest fires.

Leave No Trace: after ensuring your fire is completely out, restore the area as much as possible. Scatter any remaining ashes and return the site to its natural state. This practice not only protects the environment but also helps maintain the natural beauty of wilderness areas for others to enjoy.

Fire starting is an indispensable wilderness survival skill that requires knowledge, practice and respect for the power of fire. By mastering various fire-starting methods, understanding the materials needed and adhering to fire safety principles, you can ensure that you are prepared to build a fire in any situation. This skill not only provides warmth and safety but also offers a sense of control and comfort in the unpredictable environment of the wilderness.

Building Natural Shelters

When you find yourself in a wilderness survival situation, constructing a natural shelter is one of the most immediate and critical tasks to ensure your safety and well-being. A well-built shelter protects you from the elements, conserves body heat and provides a psychological sense of security in an otherwise harsh environment. This section will guide you through the different types of natural shelters, the techniques for constructing them and how to choose the safest and most effective location.

Types of Shelters

Natural shelters are structures made from materials readily available in your environment. They can range from simple to complex, depending on the resources at hand, the weather conditions and the time available for construction. Here are some of the most common types of natural shelters:

1. **Lean-To Shelter**: the lean-to is one of the simplest and quickest shelters to build. It consists of a single sloped roof made by leaning branches or other materials against a supporting structure, such as a tree or a ridgepole (a horizontal pole supported at both ends). The roof is then covered with layers of foliage, leaves or other debris to create insulation and protection from rain or wind. The lean-to is effective when you have limited time or materials and need to build a shelter quickly

2. **Debris Hut**: a debris hut is an excellent all-around shelter, particularly suited for colder climates. It is essentially a small, enclosed structure built from sticks, branches and a thick covering of leaves, grass or other insulating materials. The frame is usually constructed with a ridgepole supported by two forked sticks and the covering is packed densely to trap body heat. The entrance should be small to minimize heat loss and the inside can be further insulated with debris.
3. **A-Frame Shelter**: an A-frame shelter offers more protection from the elements than a lean-to. It is built by placing a long ridgepole between two sturdy supports, with branches or poles forming a steeply sloped roof on either side. Like the lean-to, the roof is covered with debris, but the A-frame's enclosed design provides better protection from wind and rain. This type of shelter is particularly useful in snowy or windy conditions.
4. **Wickiup or Tepee**: these conical shelters are made by arranging long poles in a circular pattern and tying them together at the top, forming a strong frame. The frame is then covered with layers of bark, leaves or skins to create a weather-resistant shelter. Wickiups are spacious and provide excellent protection against rain and snow, but they require more time and materials to build than simpler shelters.
5. Snow Cave: in snowy environments, a snow cave can provide excellent insulation and protection from wind. To build a snow cave, you must first dig into a deep snowdrift or pile, creating a tunnel that leads to a small chamber. The roof of the chamber should be domed to allow any melting snow to run down the walls rather than drip onto you. Snow caves are highly effective at maintaining a stable, warm interior temperature, but they must be built carefully to avoid collapse.

Construction Techniques

Building a natural shelter requires not only knowledge of different types of shelters but also practical construction techniques to ensure that the shelter is stable, weather-resistant and comfortable.

1. **Gathering Materials**: the first step in building any natural shelter is to gather materials. Choose sturdy branches for the frame and ample amounts of leaves, grass or pine needles for insulation. In some environments, you may need to use mud, bark or stones to reinforce the structure. Always collect more materials than you think you'll need, as natural shelters often require substantial insulation to be effective.
2. **Constructing the Frame**: the frame is the backbone of your shelter. Whether building a lean-to, A-frame or another structure, ensure that the frame is solid and stable. Use strong, straight branches for the main supports and secure them well into the ground or against trees. The frame should be able to support the weight of the insulating materials without collapsing.
3. **Adding Insulation**: once the frame is in place, begin layering your insulating materials. Start from the bottom and work your way up, overlapping each layer like shingles on a roof. This method helps water to run off rather than seep through. The thicker the insulation, the better your shelter will be at retaining heat and keeping out the wind and rain. For snow shelters, compact the snow to make it strong enough to support itself.

4. **Creating an Entrance**: the entrance of your shelter should be small to minimize heat loss and exposure to the elements. In shelters like debris huts, the entrance can be partially blocked with additional debris or a door made from branches and leaves. For larger shelters, such as a wickiup, consider positioning the entrance away from the prevailing wind direction.
5. **Flooring**: don't forget the floor of your shelter. A layer of leaves, grass or pine boughs will insulate you from the cold ground and provide additional comfort. In wet environments, consider building a raised platform from branches or stones to keep you off the damp ground.

Placement and Safety

Choosing the right location for your shelter is as important as the construction itself. A poorly placed shelter can expose you to unnecessary risks, while a well-placed shelter enhances safety and comfort.

1. **Site Selection**: choose a site that offers natural protection from the elements, such as near a rock face, under a large tree or in a hollow. Avoid areas prone to flooding, landslides or avalanches. Also, steer clear of insect nests, animal trails or locations with potential hazards like falling branches.
2. **Wind Direction**: position your shelter with the back facing the prevailing wind to reduce exposure to cold air and wind-driven rain or snow. This positioning helps maintain the shelter's warmth and stability.
3. **Proximity to Resources**: ensure your shelter is close to essential resources such as water, firewood and food sources. However, don't set up directly next to a water source, as these areas can become colder at night and may attract wildlife.
4. **Fire Safety**: if you plan to build a fire near your shelter, choose a site with enough space to maintain a safe distance between the fire and the shelter. Ensure that the fire is downwind to prevent sparks from igniting the shelter's materials. Clear the area around the fire of any flammable debris and never leave the fire unattended.
5. **Wildlife Considerations**: in wilderness areas, be mindful of local wildlife. Avoid setting up your shelter near animal feeding or nesting areas and store food away from your sleeping area to prevent attracting predators.

Building natural shelters is a basic survival skill that requires practice, ingenuity and an understanding of your environment. By mastering the different types of shelters, construction techniques and site selection principles, you can create a safe and effective refuge in the wilderness. These skills not only enhance your chances of survival but also empower you to face the challenges of the natural world with confidence.

Mastering wilderness survival techniques is an essential skill for anyone who ventures into the great outdoors. By understanding the importance of survival skills, knowing how to start a fire and being able to construct a natural shelter, you can significantly improve your chances of surviving in the wilderness. These skills not only provide physical protection but also help maintain morale and confidence in challenging situations. As with all survival techniques, practice and preparation are key to ensuring that you are ready to face whatever challenges the wilderness may present.

Exercise Chapter 9
Fire Starting Without Matches

Objective: learn how to start a fire using traditional methods without matches or lighters.

Materials Needed: fireboard (softwood), spindle (hardwood), bow (branch and string), handhold (stone or hardwood), flint and steel, char cloth or dry tinder, small twigs (kindling) and larger sticks (fuel wood).

1. Prepare the Fireboard and Spindle: start by selecting your fireboard and spindle. The fireboard should be made from a dry, softwood like cedar, while the spindle should be a harder wood like oak. Carve a small notch in the fireboard and create a small depression beside the notch where the spindle will sit.

2. Assemble the Bow Drill: attach a string to your bow (a sturdy, slightly curved branch). Place the spindle in the bowstring loop and position the spindle's pointed end in the fireboard's depression. Hold the top of the spindle with the handhold.

3. Start the Fire Using the Bow Drill Method: begin sawing the bow back and forth, applying steady downward pressure with the handhold to spin the spindle rapidly. As the spindle spins, friction will create heat, producing a small pile of black, hot dust at the notch in the fireboard.

4. Ignite the Ember: once you see smoke and a glowing ember forming in the dust, carefully transfer the ember to your prepared tinder bundle. Gently blow on the ember to ignite the tinder.

5. Build the Fire: once the tinder catches fire, slowly add kindling, starting with the smallest twigs and gradually moving to larger sticks. Arrange the kindling in a teepee shape to allow air to flow through the fire.

6. Test the Flint and Steel Method: as an alternative method, strike a piece of flint against a steel striker to create sparks. Direct the sparks onto char cloth or dry tinder. Once the tinder begins to glow, gently blow on it to create a flame. Add kindling to build the fire.

Deliverable: write a short report detailing your experience with the bow drill and flint and steel methods. Include any challenges you encountered, how you resolved them and photos of your successful fire starts. Reflect on the differences between the two methods and which one you found more effective.

This exercise provides hands-on experience in starting a fire without modern tools, reinforcing critical wilderness survival skills.

Chapter 10
Traps and Hunting

In the wilderness, securing food is one of the most vital tasks for survival. While gathering edible plants and fruits can sustain you for a time, hunting and trapping provide the necessary proteins and fats that are crucial for long-term survival. Understanding how to effectively hunt and set traps not only increases your chances of finding food but also conserves energy by allowing the environment to work for you. This chapter will guide you through the essential techniques and ethics of hunting, the construction and placement of traps and the basics of fishing. By mastering these skills, you can ensure a steady food supply in the wilderness.

Hunting and trapping have been fundamental human skills for millennia, allowing early humans to thrive in a variety of environments. In modern times, these skills remain invaluable, not just for those who hunt as a hobby but for anyone who ventures into the wild. This chapter emphasizes not only the practical aspects of hunting and trapping but also the ethical considerations that come with taking an animal's life. You will also learn about the necessary equipment, safety measures and different methods for catching fish, another crucial food source in many survival scenarios.

Introduction to Hunting Techniques

Hunting is both a skill and an art. It requires patience, knowledge and respect for the natural world. Whether you are hunting with a firearm, a bow or more primitive tools, the principles of hunting remain consistent: understanding the habits of your prey, using the environment to your advantage and ensuring a clean, ethical kill.

Hunting Rules and Ethics

Before delving into the techniques, it is essential to understand the ethical considerations and rules of hunting. Ethical hunting is about respecting the animals and the environment. It involves hunting only what you need, ensuring a quick and humane kill and following all local laws and regulations. In many cultures, hunting is a rite of passage and is done with a deep respect for nature. This respect includes not over-harvesting wildlife populations, avoiding hunting during breeding seasons and using every part of the animal to minimize waste.

Understanding the rules of hunting is also crucial for safety and legality. Regulations vary by region, so it's important to familiarize yourself with local laws regarding hunting seasons, permitted species and required licenses. Ethical hunters also practice safety at all times, ensuring that they do not put themselves, others or the environment at unnecessary risk.

Necessary Equipment

The equipment you need for hunting can vary depending on the type of game and the environment. However, some basic tools are essential regardless of the specifics:

- **Weapon**: this could be a rifle, bow, crossbow or even a more primitive weapon like a spear. The choice of weapon depends on the game you are hunting, local regulations and your personal skills. Ensure that your weapon is well-maintained and that you are proficient in its use before heading out.
- **Ammunition or Arrows**: always carry more ammunition or arrows than you think you will need. Malfunctions, missed shots and the need for follow-up shots can quickly deplete your supply.
- **Knife**: a good hunting knife is indispensable for field dressing game. It should be sharp, durable and easy to handle. Some hunters prefer a multi-tool that includes various blades and other useful tools.
- **Clothing**: camouflage clothing is often used to blend into the environment, but the most important aspect is that your clothing is appropriate for the weather. Layers are essential in cold climates, while moisture-wicking materials are vital in warm environments. Brightly colored vests or hats are often required by law during hunting season to prevent accidents.
- **Binoculars**: these are crucial for spotting game from a distance and assessing whether an animal is a suitable target.
- **Backpack**: a sturdy backpack is necessary for carrying your equipment, food, water and any game you harvest. It should be comfortable to wear for long periods and large enough to accommodate your gear.
- **First-Aid Kit**: hunting can be dangerous, so a first-aid kit is essential for treating injuries in the field. Your kit should include bandages, antiseptics, pain relievers and any necessary personal medications.

Hunting Safety

Safety is the top priority when hunting. Accidents can be deadly and many of them are preventable with proper precautions. Here are some key safety guidelines:

- **Always Identify Your Target**: before taking a shot, ensure that your target is clearly visible and identifiable. Never shoot at movement or sound and always be aware of what is beyond your target.
- **Know Your Weapon**: whether you're using a firearm or a bow, you must be thoroughly familiar with how your weapon operates. Practice regularly and ensure that your weapon is in good working order before heading out.
- **Communicate**: if you are hunting with others, maintain clear communication at all times. Establish a plan before you start and use signals or radios to keep in touch.
- **Wear Blaze Orange**: in many regions, wearing bright orange clothing during hunting season is mandatory. This helps other hunters see you and reduces the risk of accidental shootings.
- **Stay Aware of Your Surroundings**: always be conscious of your environment, including the location of other hunters, the terrain and weather conditions. Avoid dangerous areas like steep slopes, cliffs or areas prone to avalanches.
- **Field Dressing**: when field dressing game, be cautious of sharp knives and potential exposure to diseases. Wear gloves and handle all game meat with care to avoid contamination.

Building Traps

Trapping is an effective method of securing food in survival situations because it allows you to catch animals without expending too much energy. With the right knowledge, you can set multiple traps and increase your chances of catching small game. This section will cover the different types of traps, how to construct them and the best places to set them.

Types of Traps

There are many different types of traps, each designed to capture specific types of animals. Here are a few of the most common:

- **Snare Trap**: one of the simplest and most effective traps, a snare consists of a noose made from wire or strong cordage that tightens around the animal as it moves through it. Snares can be set on game trails or near burrows and are effective for catching small to medium-sized animals like rabbits and foxes.
- **Deadfall Trap**: a deadfall trap uses a heavy object, such as a log or rock, to crush the animal when triggered. The object is held up by a trigger mechanism, which releases when the animal takes the bait. Deadfalls are useful for catching larger animals but require more skill to set up correctly.

- **Pitfall Trap**: this is a simple trap where a hole is dug and covered with camouflage, causing the animal to fall in when it steps on the cover. Pitfall traps are effective for catching ground-dwelling animals but are labor-intensive to create.
- **Figure-Four Trap**: this is a specific type of deadfall trap with a trigger mechanism shaped like the number four. When the animal moves the bait, the trigger collapses and the heavy object falls.
- **Spring Trap**: a spring trap uses a bent sapling or branch as a spring to snatch the animal into the air when it triggers the trap. These are often used for birds or small mammals.

Materials and Construction

Building traps requires materials that are often available in the wild. Here's how to gather what you need and construct effective traps:

- **Cordage**: strong cord or wire is essential for making snares and other traps. You can use natural materials like vines or roots if necessary, but commercial wire or paracord is more reliable.
- **Trigger Mechanisms**: sticks, rocks and logs can be used to create trigger mechanisms for various traps. These need to be carefully crafted and balanced to ensure the trap functions correctly.
- **Bait**: the success of a trap often depends on the bait used. Food scraps, animal parts or even shiny objects can attract animals to the trap. Knowing the diet of the target animal is crucial for selecting the right bait.
- **Setting Up**: traps should be constructed with care, ensuring they are sensitive enough to trigger but not so delicate that they go off prematurely. Practice is essential to mastering the art of trap setting.

Trap Placement

The placement of traps is just as important as their construction. Here are some tips on where to set your traps:

- **Game Trails**: setting traps along well-used animal trails increases the likelihood of a catch. Look for tracks, droppings and signs of animals brushing against vegetation.
- **Water Sources**: animals need water, so placing traps near rivers, streams or ponds is often effective.
- **Feeding Areas**: if you find an area with abundant food sources, such as a berry patch or a field of clover, it's a good spot to set traps.
- **Natural Funnels**: look for natural features like ravines, thick brush or fallen trees that funnel animals into a specific path. Setting traps in these areas can be highly effective.
- **Burrows and Nests**: setting traps near animal burrows or nesting sites can yield results, but it requires caution to avoid scaring the animals away.

Fishing Techniques

Fishing is another essential skill in wilderness survival, providing a reliable source of protein when hunting or trapping might not be successful. This section will cover the equipment needed for fishing, various methods and how to preserve your catch.

Equipment and Preparation

Fishing requires specific tools and being prepared can make all the difference:

- **Fishing Rod and Line**: in a survival situation, you can use a makeshift rod made from a sturdy stick or even just a handline. Fishing line can be anything from commercial monofilament to sturdy string or even improvised materials like dental floss or plant fibers. The key is to ensure that your line is strong enough to hold the fish you intend to catch.
- **Hooks and Bait**: fishing hooks are small but vital pieces of equipment. If you don't have a commercial hook, you can fashion one from a thorn, bone or bent wire. Bait can range from insects and worms to bits of food or even small fish. Selecting the right bait is crucial and depends on the fish species you are targeting.
- **Floats and Weights**: floats (bobbers) help keep your bait at the desired depth, while weights (sinkers) help cast your line farther and keep the bait near the bottom. Improvised floats can be made from lightweight materials like cork or wood and weights can be fashioned from small stones or metal pieces.
- **Net or Trap**: while a fishing line is useful, nets or traps can catch multiple fish at once and are effective when you need to gather food quickly. Nets can be made from cordage and traps can be constructed from natural materials like branches or reeds.

Fishing Methods

Several methods can be employed depending on the resources available and the type of water you are fishing in:

- **Rod and Line Fishing**: the most common method, rod and line fishing, involves casting your baited hook into the water and waiting for a fish to bite. Patience and stillness are key, as sudden movements can scare fish away.
- **Handlining**: handlining is similar to rod fishing but without the rod. It involves manually pulling in the line when a fish bites. This method is useful when a rod isn't available or practical to use.
- **Spearfishing**: spearfishing involves using a spear or sharpened stick to impale fish in shallow water. This method requires skill and precision, as the water's refraction can make the fish appear at a different location than it actually is.
- **Net Fishing**: if you have access to a net, you can use it to scoop fish out of shallow water or drag it through the water to catch multiple fish at once. Nets are especially effective in rivers or streams where fish are naturally funneled into narrower areas.

- **Fishing Traps**: fishing traps can be set in rivers, streams or tidal areas to passively catch fish. These traps often use a funnel design, where fish can swim in but can't easily escape. Checking traps regularly is important to ensure your catch remains fresh and doesn't attract predators.

Fish Preservation

Once you've caught fish, preserving them is crucial to ensure they remain edible, especially if you catch more than you can consume immediately:

- **Drying**: drying fish is one of the oldest preservation methods. Clean the fish, remove the entrails and fillet it if possible. Then, hang the fish strips in a dry, sunny location with good airflow. The process can take several days but results in fish that can be stored for weeks or months.
- **Smoking**: smoking fish not only preserves it but also adds flavor. After cleaning and filleting the fish, hang it in a smokehouse or over a fire where it can be exposed to the smoke for several hours to days, depending on the thickness of the fish. The smoke helps to dehydrate the fish and adds a protective layer that keeps bacteria at bay.
- **Salting**: salting is another effective preservation method. Clean the fish and pack it in salt, ensuring that all surfaces are well-covered. The salt draws out moisture and creates an environment where bacteria cannot thrive. Salted fish can be stored for long periods, but it needs to be rehydrated or soaked before consumption.
- **Freezing**: if you are in a cold environment, freezing is the simplest way to preserve fish. Clean the fish and store it in a shaded area where it will remain frozen. However, be aware that freezing may affect the texture and flavor over time.

Mastering the techniques of hunting, trapping and fishing is crucial for anyone who spends time in the wilderness or who might find themselves in a survival situation. These skills provide not only the means to secure food but also a deeper understanding and respect for nature. Ethical hunting and trapping, careful selection and placement of traps and effective fishing methods can significantly increase your chances of survival while maintaining a balanced and respectful relationship with the environment.

By learning how to construct and place various traps, select and use the right hunting equipment and preserve your catch effectively, you ensure that you can sustain yourself and others in a survival situation. As with all survival skills, practice and preparation are key. The more familiar you are with these techniques, the more confident and capable you will be when the need arises.

Exercise Chapter 10
Setting Up a Basic Snares for Small Game

Objective: learn how to set up a basic snare to catch small game in a survival situation.

1. Materials Needed: strong wire (24-gauge or similar), paracord or sturdy string, a knife or multitool, small sticks or branches (for support) and gloves (for scent control).

2. Select a Location: identify an area where small game, such as rabbits or squirrels, frequently travel. Look for well-worn paths, tracks, droppings or areas near food sources. Ideal spots are near natural funnels like dense brush or along the edges of clearings.

3. Prepare the Snare Wire: cut a piece of wire approximately 18 to 24 inches long. Form a small loop (about the size of a pencil) at one end of the wire by twisting it around itself several times. This loop will be the locking mechanism for the snare.

4. Create the Noose: thread the other end of the wire through the small loop to create a noose. The noose should be large enough for the animal's head to pass through but tight enough to trap the animal as it tries to escape. For rabbits, a 3- to 4-inch diameter loop is ideal, set about 3 inches off the ground.

5. Set Up the Snare: use two sturdy sticks or branches as supports on either side of the trail. Firmly plant these sticks in the ground. Attach the snare wire to one support stick or a nearby tree/shrub. Position the noose in the middle of the trail, directly in the animal's path.

6. Camouflage the Snare: lightly cover the wire with grass, leaves or other natural materials to blend it into the environment without obstructing the noose. This step helps avoid alarming the animal as it approaches.

7. Check and Maintain: regularly check the snare to see if it has caught anything. Avoid leaving snares unattended for long periods to minimize the animal's suffering and prevent scavengers from stealing the catch. If you find the snare disturbed but empty, try adjusting the height or placement.

Deliverable: write a brief report on setting up the snare. Include any challenges, solutions and photos of your setup. Reflect on the effectiveness of the placement and any future improvements.

This exercise provides practical experience in constructing and setting a basic snare, an essential skill for survival situations where hunting small game is necessary.

Chapter 11
Building Emergency Structures

Here's the revised introduction, maintaining the structure you provided: In any emergency scenario, the ability to swiftly construct reliable structures is an invaluable skill. Whether it's assembling a durable shed to safeguard essential supplies, creating a temporary shelter to provide immediate protection or building a storage space to secure vital tools and resources, understanding how to build sturdy, functional structures is crucial.

These structures go beyond merely offering shelter or storage; they form the backbone of your survival strategy, ensuring safety and reliability when it matters most. In this chapter, we will delve into the principles of emergency structure building, offering insights into both permanent and temporary solutions, with practical guidance on planning, design and construction.

Introduction to Structure Building

Building emergency structures is a key aspect of preparedness, vital for creating secure, functional spaces that can support you and your loved ones during critical times. Whether you're preparing for natural disasters like hurricanes, floods or earthquakes or anticipating other unforeseen disruptions, knowing how to construct essential structures significantly enhances your resilience. These structures are integral to your emergency plan, providing protection from the elements, safe storage for supplies and a secure base from which to manage your survival efforts.

The necessity for emergency structures can arise in various situations, each demanding specific types of buildings. Generally, these structures fall into three categories: storage buildings, sheds and temporary shelters. Each type serves a unique purpose and requires tailored construction techniques to ensure it withstands the challenges it may face. Understanding these different types and their specific functions is the first step in effective emergency structure planning.

Types of Necessary Structures

Understanding the types of structures you might need is crucial for effective emergency planning. Storage buildings and sheds are typically more permanent and are designed to securely store supplies, tools and other essential items. These structures must be built to endure harsh conditions, ensuring the protection of their contents. Proper internal organization is also key, as it allows you to access necessary items quickly when time is of the essence.

Temporary shelters, on the other hand, are designed for short-term use and need to be quickly assembled. These shelters are crucial in providing immediate protection from the elements, whether you're dealing with the aftermath of a disaster or setting up a temporary base of operations in an unfamiliar environment. The ability to construct a reliable temporary shelter can mean the difference between safety and exposure in situations where permanent structures are not available or feasible.

Planning and Design

Planning and design are integral components of building effective emergency structures. A well-conceived plan ensures that your structure will meet your specific needs and be resilient enough to withstand the conditions it may encounter. This planning process involves several key considerations:

- **Location**: choosing the right location is crucial. You'll need to consider factors such as proximity to resources (like water and firewood), the likelihood of natural hazards (like flooding or landslides) and accessibility. The location should offer natural protection from the elements while being strategically placed to serve its intended purpose.
- **Size and Layout**: the size of the structure should be proportional to its intended use. A storage shed, for example, needs to be large enough to hold all necessary supplies but compact enough to maintain structural integrity. Similarly, a temporary shelter should be spacious enough to accommodate occupants comfortably while being easy to heat and cool.
- **Materials and Tools**: the materials and tools you choose will largely depend on the type of structure you're building and the environment in which you're working. For instance, wood is a versatile and widely available material that can be used for both sheds and shelters. Metal might be preferred for storage buildings due to its durability and resistance to pests. Tools can range from basic hand tools like hammers and saws to more advanced equipment like power drills, depending on the complexity of the build.

- **Environmental Considerations**: understanding the specific risks and challenges of your environment is vital. This includes weather patterns, available resources and potential hazards such as high winds or seismic activity. For example, in hurricane-prone areas, you might prioritize building a wind-resistant structure, while in colder climates, insulation and heat retention become key concerns.

Materials and Tools

Selecting the appropriate materials and tools is essential for constructing durable and functional emergency structures. Your choice of materials should be guided by the structure's intended use, the environmental conditions and the resources available to you. Common materials include:

- **Wood**: versatile, easy to work with and widely available, wood is ideal for constructing both sheds and temporary shelters. However, it must be treated to resist rot and insects, especially in humid environments.
- **Metal**: often used for storage buildings, metal provides excellent durability and protection against pests. It is, however, more challenging to work with and requires specific tools for cutting and assembly.
- **Plastics and Composites**: these materials are lightweight and resistant to weathering, making them suitable for certain types of shelters or as components in more complex structures.

The tools you use will depend on the materials and the complexity of the structure. Basic hand tools like hammers, saws and screwdrivers are essential for most builds, while more advanced tools like power drills, nail guns and saws may be necessary for more sophisticated constructions. Having a well-equipped toolkit is crucial for ensuring that your structure is built to last.

In summary, building emergency structures requires careful planning, the right materials and appropriate tools. Whether you're constructing a simple shed or a complex temporary shelter, understanding the principles of design and construction will help you create structures that are not only functional but also resilient in the face of adversity.

Sheds and Storage Buildings

When preparing for emergencies, sheds and storage buildings play a crucial role in keeping your essential supplies, tools and equipment safe and easily accessible. These structures serve as the backbone of your preparedness strategy, providing secure storage for everything from food and water supplies to tools, fuel and other critical resources. In this section, we will explore key considerations for building a shed, organizing its interior efficiently and maintaining it to ensure long-term functionality and durability.

Building a Shed

Constructing a shed is a practical project that requires careful planning, the right materials and a methodical approach to ensure the final structure is both functional and resilient. A well-built shed should be sturdy enough to withstand harsh weather conditions, secure enough to protect valuable supplies and spacious enough to store everything you need for your emergency preparedness.

1. Planning the Shed: the first step in building a shed is to determine its purpose and size. Ask yourself what you intend to store in the shed, as this will influence the size and layout. For instance, if you need to store large items like gardening tools, emergency generators or bulk food supplies, you'll require a larger structure with ample floor space. Conversely, a smaller shed might suffice for storing more compact items like hand tools, camping gear or non-perishable food.

2. Selecting the Location: choosing the right location for your shed is critical. The site should be level, well-drained and easily accessible. Avoid areas prone to flooding or those with poor soil stability, as these can compromise the shed's foundation. Additionally, consider the proximity of the shed to your home or main living area, convenience is key during an emergency, so the shed should be easily reachable.

3. Foundation and Flooring: a solid foundation is essential for a durable shed. Common foundation options include concrete slabs, gravel beds or treated wooden beams. The foundation must be level and strong enough to support the weight of the shed and its contents. For the flooring, pressure-treated plywood or concrete are durable options that provide a stable base and help protect against moisture.

4. Framing and Structure: the framing of the shed is crucial for its overall strength and durability. Use treated lumber to resist rot and insect damage and ensure that the framing is properly aligned and securely fastened. The walls and roof should be built with weather-resistant materials, such as treated wood, metal panels or composite materials, to protect against the elements. For the roof, consider a sloped design to facilitate water runoff and prevent pooling.

5. Ventilation and Insulation: proper ventilation is important to prevent moisture buildup inside the shed, which can lead to mold, mildew and deterioration of stored items. Install vents near the roofline to allow air circulation. If you plan to store temperature-sensitive items, consider insulating the shed to protect against extreme heat or cold. Insulation materials such as foam boards or fiberglass can help maintain a more stable interior temperature.

Internal Organization

Once the shed is built organizing the interior efficiently is key to making the most of the available space and ensuring that you can quickly locate items in an emergency.

1. Shelving and Storage Solutions: install sturdy shelves along the walls to maximize vertical space. Adjustable shelving systems are ideal, as they allow you to customize the layout to accommodate items of varying sizes. For smaller items, use storage bins or drawers that can be labeled for easy identification. Pegboards are another useful addition, providing a convenient way to hang tools and other frequently used items.

2. Space Management: divide the shed into zones based on the types of items stored. For example, dedicate one area to emergency supplies like water, food and medical kits, while another area can be used for tools and equipment. This zoning approach helps maintain order and ensures that essential items are easily accessible when needed.

3. Security Features: given that sheds often contain valuable and essential items, security is a top priority. Install a robust locking system on the shed doors to deter unauthorized access. If possible, add a security light or camera to monitor the shed, especially if it is located in a more secluded part of your property. Additionally, consider securing high-value items, such as generators or fuel, with chains or padlocks inside the shed.

Structure Maintenance

To ensure that your shed remains in good condition and continues to serve its purpose, regular maintenance is necessary. A well-maintained shed not only protects its contents but also extends the lifespan of the structure itself.

1. Regular Inspections: conduct routine inspections of the shed, focusing on the roof, walls, foundation and door. Look for signs of wear and tear, such as leaks, cracks or rot and address any issues promptly to prevent further damage. Pay special attention to the roof, as it is the most exposed part of the shed and is susceptible to damage from weather conditions.

2. Pest Control: pests such as rodents, insects and birds can cause significant damage to the shed and its contents. Seal any gaps or holes in the structure to prevent pests from entering and use traps or deterrents as necessary. Regularly check for signs of pest activity, such as droppings or chewed materials and take action to eliminate any infestations.

3. Weatherproofing: over time, weatherproofing materials can degrade, leaving the shed vulnerable to moisture and other environmental factors. Reapply sealants or paint as needed to maintain a protective barrier against the elements. Ensure that the roof remains watertight by checking for loose or damaged shingles and making repairs as needed.

4. Cleaning and Organization: keep the interior of the shed clean and organized to prevent clutter and ensure that items are stored properly. Periodically declutter the space, removing any items that are no longer needed or that have expired. This not only makes it easier to find what you need but also reduces the risk of pests and improves overall safety.

Sheds and storage buildings are indispensable components of a well-rounded emergency preparedness plan. By carefully planning and constructing these structures organizing the interior effectively and performing regular maintenance, you can ensure that your supplies and equipment are safely stored and readily accessible when you need them most. These structures provide peace of mind, knowing that you are prepared to face whatever challenges may arise.

Temporary Shelters

Temporary shelters are essential components of any emergency preparedness plan, designed to provide immediate protection and comfort when permanent structures are either unavailable or unsuitable. These shelters are particularly crucial in situations where time is of the essence, such as during natural disasters, evacuations or when traveling through remote areas. The ability to quickly assemble and disassemble these shelters, coupled with their portability and adaptability, makes them invaluable for survival and emergency scenarios. In this section, we will explore the various types of temporary shelters, the materials and techniques used in their construction and important considerations for ensuring safety and comfort.

Building Portable Shelters

Portable shelters are designed for rapid deployment and ease of transport, making them ideal for short-term use in emergency situations. They can range from commercially manufactured tents to improvised structures built using materials found in the environment. The key to effective portable shelters is their simplicity and efficiency, allowing for quick setup with minimal tools and resources.

Portable shelters come in many forms, each suited to different conditions and needs:
- **Tents**: tents are the most common type of portable shelter, widely used in camping, disaster response and survival situations. They are typically made from durable, weather-resistant materials such as nylon or polyester and can be set up quickly using poles and stakes. Tents vary in size from small, one-person designs to larger family tents and often include features such as waterproof rainflys, ventilation windows and integrated ground sheets.
- **Tarps and Bivouacs**: a tarp or bivouac (bivy) sack is a more minimalist shelter option, ideal for situations where weight and packability are critical. Tarps can be rigged in various configurations using ropes or paracord, providing a lightweight, flexible shelter that can be adapted to different environments. A bivy sack, on the other hand, is a waterproof cover designed to enclose a sleeping bag, offering protection from the elements without the bulk of a full tent.
- **Pop-Up Shelters**: these are lightweight, pre-assembled shelters that "pop up" into shape when unfolded. Pop-up shelters are incredibly convenient, requiring minimal setup time and are often used for short-term emergencies, such as sudden rainstorms or as temporary medical tents. However, they may not offer the same level of durability or weather protection as more robust shelter types.

Improvised Shelters: In situations where pre-made shelters are unavailable, improvisation becomes necessary. Improvised shelters can be made using natural materials such as branches, leaves and rocks or by repurposing available items like tarps, blankets and plastic sheeting. While these shelters require more skill and creativity to construct, they can be highly effective in providing immediate protection.

Materials and Techniques

The materials and construction techniques for temporary shelters are chosen based on the specific needs of the situation, including the expected weather conditions, available resources and the duration of use. The goal is to create a shelter that is both effective and efficient, using materials that are strong, lightweight and easy to work with.

Common Materials:
- **Fabrics**: high-quality, weather-resistant fabrics like ripstop nylon or polyester are commonly used in tents and tarps. These materials are lightweight, durable and provide good protection against wind, rain and UV rays. For improvised shelters, blankets, plastic sheeting or even large garbage bags can be used to create a barrier against the elements.
- **Poles and Supports**: tent poles are typically made from lightweight materials like aluminum, fiberglass or carbon fiber, which offer a good balance of strength and portability. For improvised shelters, branches, sticks or trekking poles can serve as structural supports. These supports are crucial for maintaining the shelter's shape and stability.
- **Cordage**: rope, paracord or guy lines are essential for securing shelters, especially when using tarps or improvised materials. Cordage is used to anchor the shelter to the ground, tie off sections to trees or other supports and create tension to keep the shelter taut and weather-resistant.
- **Insulation and Ground Cover**: to improve comfort and protection from the cold, insulation materials such as foam pads, emergency blankets or leaves can be used inside the shelter. Ground cover is also important to prevent moisture from seeping in, tarps, ground sheets or even a thick layer of pine needles can serve this purpose.

Construction Techniques:
- **Site Selection**: the first step in constructing a temporary shelter is choosing an appropriate site. The site should be flat, well-drained and provide natural protection from the wind, such as behind a hill or under a canopy of trees. Avoid low-lying areas where water might pool and sites that are exposed to falling branches or rocks.
- **Anchoring**: proper anchoring is essential to keep the shelter secure, especially in windy conditions. Stakes, rocks or logs can be used to anchor the corners of the shelter. If stakes are not available, natural features like trees or large boulders can be used to tie off the shelter. For tarps, creating a ridgeline between two trees is a common technique to provide a stable anchor point.
- **Layering for Insulation**: in colder conditions, layering materials is key to retaining warmth. For example, in a tent, placing an emergency blanket over the interior can reflect heat back inside, while adding leaves or other insulating materials between the ground and the sleeping area can reduce heat loss.
- **Waterproofing**: ensuring the shelter is waterproof is crucial in wet conditions. Tents often come with built-in waterproofing, but for improvised shelters, using a tarp or plastic sheeting as a roof can help keep the interior dry. Slope the roof of the shelter to allow rainwater to run off and dig small trenches around the shelter's perimeter to divert water away.

Safety and Comfort

When setting up temporary shelters, safety and comfort are paramount. A poorly constructed or improperly placed shelter can lead to exposure, injury or other hazards, so it's essential to consider both the physical structure and the environment around it.

Safety Considerations:
- **Structural Integrity**: ensure the shelter is stable and can withstand the expected weather conditions. Regularly check for any signs of sagging or loosening of ropes and make adjustments as necessary.
- **Fire Safety**: if you plan to build a fire near the shelter, place it at a safe distance to prevent sparks from reaching the shelter materials. Additionally, ensure that the shelter is well-ventilated to avoid the buildup of smoke or fumes inside.
- **Wildlife Protection**: in areas where wildlife is a concern, store food away from the shelter and use proper food storage techniques, such as hanging food bags from trees. Keeping the shelter clean and free of food debris will help reduce the risk of attracting animals.

Ensuring Comfort:
- **Ventilation**: adequate ventilation is important to prevent condensation inside the shelter, which can lead to discomfort and hypothermia. Tents often come with built-in vents or mesh windows, while improvised shelters can be designed with open sections that allow air to flow through.
- **Temperature Regulation**: depending on the environment, you may need to focus on either retaining heat or keeping cool. In cold weather, insulating the shelter and blocking wind entry points is crucial. In hot climates, creating shade and allowing for airflow is more important.
- **Sleeping Arrangements**: ensure that the sleeping area is dry and comfortable. Use a sleeping pad or a layer of insulating material to cushion against the ground. In a tent, position sleeping bags or mats on higher ground within the shelter to stay dry in case of rain.

Temporary shelters, while intended for short-term use, are critical for immediate protection in emergencies. By understanding the types of shelters available, mastering the materials and techniques required for their construction and prioritizing safety and comfort, you can create an effective temporary refuge that will keep you safe and secure until more permanent solutions are available. Whether you're preparing for a natural disaster, planning an outdoor adventure or simply looking to improve your survival skills, the ability to build a temporary shelter is an invaluable skill that can make a significant difference in your preparedness and resilience.

In conclusion, mastering the construction of emergency structures is an invaluable skill in any preparedness plan. Whether you're building a permanent shed to store essential supplies or a temporary shelter to protect against the elements, the knowledge and techniques covered in this chapter will equip you to create safe, functional spaces that enhance your resilience. These skills ensure that, no matter the circumstances, you are prepared to protect yourself and your resources effectively.

Exercise Chapter 11
Building a Storage Shed for Emergency Supplies

Objective: learn how to build a storage shed specifically designed for storing emergency supplies, focusing on durability organization and accessibility.

Materials Needed: pressure-treated wood (for framing and flooring), plywood (for walls and roofing), roofing materials (such as shingles or metal sheets), nails and screws, hinges and a sturdy lock, shelving units or materials to build shelves, weatherproofing sealant and basic tools (hammer, saw, drill, level).

1. Plan the Shed Design: determine the shed size based on the supplies you'll store. Sketch a simple floor plan, considering location, size and ease of access.

2. Prepare the Foundation: clear and level the site. Use gravel, a concrete slab or treated wooden beams as a foundation, ensuring it is solid and level.

3. Construct the Frame: build the floor frame using pressure-treated wood, followed by the walls. Ensure the structure is square and plumb and add roof trusses to support the roof.

4. Install the Roof: attach plywood sheets to the roof trusses, then apply the roofing material. Ensure the roof has a slight pitch for rain runoff.

5. Build and Install the Door: construct a sturdy door, attach it with heavy-duty hinges and install a reliable lock. Ensure the door allows easy access.

6. Add Shelving and Storage: install shelves or use labeled bins for organization. Ensure frequently needed items are easily accessible.

7. Weatherproof and Seal: apply weatherproofing sealant to protect the shed from moisture. Ensure proper ventilation to prevent condensation.

Deliverable: write a short report detailing your experience building the shed. Include challenges, solutions and photos of your completed shed. Reflect on the organization and how well it meets your storage needs.

This exercise provides hands-on experience in constructing a storage shed, an essential part of emergency preparedness.

Chapter 12
Long-Term Food

In any plan for long-term self-sufficiency, securing a reliable and sustainable food source is paramount. While short-term food production focuses on immediate needs and quick harvests, long-term food production requires a more strategic approach, emphasizing sustainability, resilience and the ability to continuously provide nourishment over extended periods. This is not just about growing food; it is about creating a system that can endure the challenges of changing seasons, climate variability and unforeseen disruptions. Long-term food production integrates principles of ecology, sustainability and traditional farming practices to create a resilient food supply that can support you and your family indefinitely.

As you look to establish or enhance your long-term food production system, it's important to consider methods that go beyond the basics of gardening and small-scale animal husbandry. This chapter will explore advanced techniques that can significantly boost the sustainability and productivity of your food system. From agroforestry and perennial farming systems to the crucial practice of seed saving and crop rotation, these strategies are designed to create a balanced, self-sustaining food production environment that reduces dependency on external inputs and enhances the resilience of your homestead.

Main Aspects of Long-Term Food Production

Long-term food production requires a comprehensive approach that integrates various practices aimed at sustainability, resilience and ecological balance. These aspects go beyond immediate food production, focusing instead on creating a system that can sustain itself and provide for human needs over extended periods, even under challenging conditions. Here are the key aspects that form the foundation of a successful long-term food production strategy:

Sustainability

Sustainability is at the core of long-term food production. It involves creating a food system that can continue to function effectively without depleting natural resources, harming the environment or relying excessively on external inputs. Sustainability in food production is achieved through several key practices:

- **Soil Health**: maintaining and improving soil health is critical for sustainable food production. Practices such as crop rotation, cover cropping and the use of organic matter (like compost and manure) help to build and preserve soil fertility. Healthy soil supports robust plant growth, reduces the need for chemical fertilizers and enhances the soil's ability to retain water, thus reducing the need for irrigation.
- **Water Management**: efficient use of water is essential, particularly in regions where water is scarce. Techniques such as rainwater harvesting, drip irrigation and the use of drought-resistant crop varieties can help ensure that water resources are used wisely and that food production can continue even during periods of drought.
- **Energy Efficiency**: reducing the energy inputs required for food production is another important aspect of sustainability. This can be achieved through the use of renewable energy sources, such as solar or wind power and by adopting low-energy farming practices, such as no-till farming and the use of hand tools where appropriate.
- **Biodiversity**: a diverse agricultural system is more resilient and sustainable. By cultivating a variety of crops and raising different species of animals, farmers can reduce the risk of pests and diseases, improve soil health and create a more stable and productive ecosystem. Biodiversity also provides a wider range of foods, improving diet diversity and nutritional security.

Resilience

Resilience refers to the ability of a food production system to withstand and recover from shocks, such as extreme weather events, pest infestations or economic disruptions. Building resilience into your food production system is essential for ensuring its long-term viability. Key practices include:

- **Diversification**: growing a wide range of crops and raising different types of livestock can help protect against the failure of any single crop or animal. This diversity ensures that even if one part of the system is compromised, others can continue to provide food.
- **Agroecological Practices**: incorporating principles of agroecology – such as integrating crops and livestock, using natural pest control methods and preserving natural habitats – can enhance the resilience of a farming system. These practices help to create a balanced ecosystem that is better able to cope with environmental stresses.

- **Adaptability**: a resilient food production system is one that can adapt to changing conditions. This might involve selecting crop varieties that are more tolerant of heat, drought or pests or adjusting planting and harvesting schedules in response to changing weather patterns.
- **Self-Reliance**: reducing dependence on external inputs – such as commercial seeds, fertilizers and pesticides – enhances resilience by making the system more self-sufficient. Practices like seed saving, composting and on-farm feed production can reduce costs and increase autonomy.

Resource Management

Effective resource management is critical for sustaining long-term food production. This involves not only conserving and efficiently using natural resources like soil, water and energy but also managing human and financial resources wisely. Key aspects of resource management include:

- **Soil Conservation**: preventing soil erosion and degradation is essential for maintaining long-term productivity. Techniques such as contour plowing, terracing and the use of perennial crops help to protect soil from erosion. Additionally, maintaining soil cover through cover cropping or mulching reduces erosion and builds organic matter.
- **Water Conservation**: water is a finite resource and its careful management is crucial for sustainable agriculture. Efficient irrigation systems, such as drip or sprinkler irrigation, minimize water waste. Collecting and storing rainwater, using gray water and implementing water-saving technologies also contribute to more sustainable water use.
- **Nutrient Cycling**: recycling nutrients within the farm system – such as composting plant residues, using animal manure and growing nitrogen-fixing crops – reduces the need for external fertilizers and helps maintain soil fertility. This closed-loop approach minimizes waste and keeps the farming system self-sufficient.
- **Labor and Time Management**: efficiently managing labor and time is vital for the success of a long-term food production system. This includes planning work schedules to align with seasonal cycles, optimizing the use of labor-saving tools and machinery and ensuring that all tasks are completed in a timely manner to avoid losses or inefficiencies.

Ecological Balance

Maintaining ecological balance is essential for the health and productivity of a long-term food production system. This involves working with, rather than against, natural processes to create a farming system that is sustainable and self-regulating. Key practices include:

- **Integrated Pest Management (IPM)**: IPM combines biological, cultural, mechanical and chemical methods to control pests in a way that minimizes harm to the environment and human health. By encouraging natural predators, using pest-resistant crop varieties and employing crop rotation, farmers can reduce the need for chemical pesticides and create a more balanced ecosystem.

- **Habitat Preservation**: preserving natural habitats, such as woodlands, wetlands and hedgerows, within or around the farm helps to support biodiversity and ecological balance. These habitats provide homes for beneficial insects, birds and other wildlife that contribute to pest control, pollination and other ecosystem services.
- **Pollinator Support**: ensuring the health of pollinators, such as bees, butterflies and other insects, is crucial for many crops. Planting a variety of flowering plants, avoiding the use of harmful pesticides and providing nesting sites can help support pollinator populations and enhance crop yields.

Concluding, the main aspects of long-term food production revolve around creating a sustainable, resilient and ecologically balanced system. By focusing on these key principles, you can develop a food production system that not only meets your immediate needs but also preserves the health of your land and resources for future generations.

Agroforestry and Perennial Systems

Agroforestry and perennial systems are sustainable agricultural practices that integrate trees, shrubs and other perennial plants into farming systems. These methods offer numerous benefits for long-term food production, including increased biodiversity, improved soil health and enhanced resilience against environmental changes. By combining agriculture and forestry, agroforestry systems mimic natural ecosystems, creating a more sustainable and productive farming environment.

The Principles of Agroforestry

Agroforestry is based on the principle of integrating trees and shrubs into agricultural landscapes. This approach offers several advantages, such as enhanced soil fertility, better water retention and increased biodiversity. Agroforestry systems can take many forms, depending on the specific goals of the farm and the local environment. The key components of agroforestry include:

- **Trees and Shrubs**: these are central to agroforestry systems. Trees provide shade, windbreaks and habitat for wildlife, while their roots help to stabilize soil and improve water infiltration. Shrubs can act as living fences, provide additional habitat and contribute to soil fertility through nitrogen fixation.
- **Diverse Planting**: agroforestry systems often involve the cultivation of a variety of crops alongside trees and shrubs. This diversity can reduce the risk of crop failure due to pests or diseases and can also enhance the resilience of the farming system to environmental stresses such as drought or flooding.
- **Integrated Livestock**: in some agroforestry systems, livestock is integrated into the landscape. Animals can graze on undergrowth or crop residues, helping to manage vegetation and fertilize the soil with their manure. This practice can improve soil health and reduce the need for external inputs like synthetic fertilizers.
- **Multifunctionality**: agroforestry systems are designed to be multifunctional, providing multiple benefits such as food, fuel, timber and ecological services. This approach increases the sustainability and economic viability of farming operations.

Benefits of Agroforestry

Agroforestry offers numerous benefits that contribute to the sustainability and productivity of long-term food production systems:

- **Soil Health**: trees and shrubs in agroforestry systems contribute to soil health by reducing erosion, increasing organic matter and improving nutrient cycling. The deep roots of trees bring up nutrients from the subsoil, making them available to other plants.
- **Water Management**: agroforestry systems enhance water management by improving soil structure and increasing water infiltration. Trees act as natural sponges, reducing surface runoff and helping to maintain groundwater levels. This is particularly important in areas prone to drought or irregular rainfall.
- **Biodiversity**: by integrating different plant species and creating diverse habitats, agroforestry promotes biodiversity. This diversity helps to control pests and diseases naturally, reducing the need for chemical inputs. It also supports a wide range of wildlife, contributing to overall ecosystem health.
- **Climate Resilience**: agroforestry systems are more resilient to climate change than monoculture systems. The diversity of species and the presence of perennial plants buffer the effects of extreme weather events, such as droughts or storms. Additionally, trees sequester carbon, helping to mitigate climate change by capturing and storing atmospheric CO_2.
- **Economic Stability**: agroforestry provides multiple income streams, reducing the financial risk for farmers. The production of timber, fruits, nuts and other non-timber forest products can complement traditional crop production, offering economic stability even in the face of fluctuating market conditions.

Perennial Systems in Agriculture

Perennial systems involve the cultivation of plants that live and produce for several years without needing to be replanted each season. Perennials include fruit and nut trees, berry bushes and perennial vegetables like asparagus or rhubarb. These systems offer several advantages for long-term food production:

- **Low Maintenance**: perennial plants require less maintenance than annual crops because they do not need to be replanted every year. This reduces labor and input costs, making perennial systems more sustainable and cost-effective over time.
- **Soil Conservation**: perennials have extensive root systems that help to hold soil in place, reducing erosion. Their deep roots also access nutrients and water from deeper soil layers, making them more resilient to drought and nutrient-poor soils.
- **Biodiversity and Habitat Creation**: perennial systems support biodiversity by providing continuous cover and habitat for a wide range of organisms. The presence of perennial plants can attract beneficial insects, birds and other wildlife, creating a balanced ecosystem that supports pest control and pollination.

- **Sustainable Harvesting**: perennial plants can be harvested over many years, providing a stable source of food or income without the need for replanting. This sustainability makes perennials ideal for long-term food production, especially in areas where annual cropping may be challenging.
- **Carbon Sequestration**: like trees in agroforestry systems, perennial plants sequester carbon, contributing to climate change mitigation. Their long lifespan allows them to capture and store carbon over extended periods, helping to reduce the overall carbon footprint of the farming system.

Implementing Agroforestry and Perennial Systems

Implementing agroforestry and perennial systems requires careful planning and a long-term perspective. Farmers must consider the specific conditions of their land, including soil type, climate and existing vegetation. Some steps to successfully implement these systems include:

1. **Site Assessment**: understanding the land's characteristics is crucial for selecting appropriate tree and perennial species. Soil tests, climate data and an assessment of existing vegetation help in making informed decisions.
2. **Designing the System**: the design of an agroforestry or perennial system should consider the spatial arrangement of plants, the interaction between different species and the overall goals of the farm. This might include planning for windbreaks, shelterbelts or integrating livestock.
3. **Selecting Species**: choosing the right species is key to the success of the system. Native species are often preferred because they are well adapted to local conditions. The selection should also consider the intended uses of the plants, such as food production, timber or soil improvement.
4. **Management and Maintenance**: once established, these systems require ongoing management to ensure their productivity and health. This includes regular monitoring, pruning, thinning and, if necessary, pest and disease control.
5. **Education and Training**: implementing agroforestry and perennial systems may require new skills and knowledge. Farmers should seek out education and training opportunities to learn about these systems and how to manage them effectively.

Agroforestry and perennial systems are integral components of sustainable, long-term food production. They offer numerous benefits, including improved soil health, enhanced biodiversity, better water management and increased resilience to climate change. By integrating these systems into their farming practices, producers can create more sustainable and productive agricultural landscapes that are capable of supporting human needs for generations to come.

Seed Saving and Crop Rotation for Sustainability

Seed saving and crop rotation are two foundational practices in sustainable agriculture that ensure the long-term health and productivity of farming systems. These techniques not only help maintain soil fertility and reduce the need for chemical inputs but also preserve genetic diversity and enhance the resilience of crops against pests and diseases. In this section, we will delve into the importance of seed saving and crop rotation, the methods involved and how these practices contribute to a more sustainable agricultural system.

The Importance of Seed Saving

Seed saving is the practice of collecting and storing seeds from crops for future planting. This age-old technique has been a cornerstone of agriculture for thousands of years, allowing farmers to adapt crops to local conditions and pass down valuable traits from one generation to the next. In modern times, seed saving has gained renewed importance as a way to preserve heirloom varieties, maintain genetic diversity and promote food sovereignty.

Preservation of Heirloom Varieties: heirloom seeds are open-pollinated varieties that have been passed down through generations. These seeds are valued for their unique flavors, resilience and adaptability to local growing conditions. By saving and replanting heirloom seeds, farmers and gardeners help preserve these varieties, which might otherwise be lost due to the dominance of commercially bred hybrids and genetically modified organisms (GMOs).

Genetic Diversity: saving seeds contributes to the maintenance of genetic diversity within crops. Genetic diversity is crucial for the resilience of agricultural systems, as it provides a pool of traits that can help crops withstand pests, diseases and changing environmental conditions. A diverse gene pool also reduces the risk of crop failure, as different varieties may react differently to stress factors, ensuring that at least some crops survive.

Local Adaptation: over time, seeds saved from crops that thrive in a particular environment become better adapted to local conditions. This adaptation includes resistance to local pests and diseases, tolerance to specific soil types and suitability to the local climate. As a result, locally adapted seeds can outperform commercial varieties that may not be suited to the unique challenges of a particular area.

Food Sovereignty: seed saving empowers farmers and communities to maintain control over their food production. By saving and sharing seeds, farmers reduce their dependence on commercial seed companies and promote a system of agriculture that is more decentralized and community-driven. This autonomy is especially important in regions where access to seeds and other agricultural inputs is limited or unreliable.

Methods of Seed Saving

Seed saving requires knowledge of plant biology, careful selection of seeds and proper storage techniques to ensure the viability and integrity of the seeds. The process varies depending on the type of plant, but the general steps include:

- **Selection of Parent Plants**: the first step in seed saving is to choose the best plants from which to collect seeds. These plants should be healthy, productive and exhibit the desirable traits you wish to preserve.

For example, if you want to save seeds from a tomato plant, select fruits that are well-formed, flavorful and free from disease.
- **Pollination Considerations**: some plants are self-pollinating, meaning they can produce viable seeds without cross-pollination from another plant. Examples include tomatoes, beans and peas. These plants are relatively easy to save seeds from, as they maintain their genetic integrity across generations. In contrast, cross-pollinating plants, such as corn, squash and cucumbers, require careful management to avoid unwanted cross-breeding. This might involve isolating plants or hand-pollinating flowers to ensure that the seeds remain true to type.
- **Seed Harvesting**: once the seeds are mature, they must be harvested at the right time. For many vegetables, this means allowing the fruit or seed pod to fully ripen on the plant. For example, beans should be left on the vine until the pods are dry and brown. After harvesting, seeds need to be thoroughly dried to prevent mold and decay. This can be done by spreading them out in a well-ventilated area away from direct sunlight.
- **Cleaning and Storage**: after drying, seeds should be cleaned to remove any debris, pulp or chaff. This can be done by hand or using sieves and screens. Once clean, seeds should be stored in a cool, dry and dark place, ideally in airtight containers to protect them from moisture and pests. Properly stored seeds can remain viable for several years, although germination rates may decline over time.
- **Labeling and Record Keeping**: it's essential to label saved seeds with the plant variety, date of harvest and any other relevant information. Keeping records of the performance of each variety, such as yield, flavor and disease resistance, can help you make informed decisions about which seeds to save and replant in future seasons.

The Role of Crop Rotation in Sustainable Agriculture

Crop rotation is the practice of growing different types of crops in the same area in sequential seasons. This method is a fundamental aspect of sustainable agriculture, offering numerous benefits for soil health, pest and disease management and overall farm productivity.

Soil Fertility Management: different crops have varying nutrient requirements and root structures. By rotating crops, farmers can optimize nutrient use and prevent the depletion of soil nutrients. For example, legumes such as beans and peas fix nitrogen in the soil, enriching it for subsequent crops that require higher nitrogen levels, like corn or wheat. This natural nutrient cycling reduces the need for synthetic fertilizers, promoting a more sustainable and environmentally friendly farming system.

Pest and Disease Control: crop rotation is an effective strategy for managing pests and diseases. Many pests and pathogens are specific to particular plant families and can build up in the soil if the same crop is grown repeatedly in the same location. By rotating crops with different susceptibilities, farmers can break the life cycles of these pests and diseases, reducing their populations and the likelihood of outbreaks.

Weed Suppression: rotating crops with different growth habits and planting schedules can help suppress weed populations. For example, planting cover crops or fast-growing species can outcompete weeds for light, water and nutrients. Additionally, some crops, like buckwheat, have allelopathic properties, meaning they release natural compounds that inhibit weed growth.

Soil Structure Improvement: different crops have varying root structures, which can positively impact soil structure. Deep-rooted crops like alfalfa or sunflowers can break up compacted soil layers, improving water infiltration and root penetration for subsequent crops. This enhancement of soil structure contributes to better soil health and increased productivity over the long term.

Diverse Crop Rotations: implementing a diverse crop rotation plan is key to maximizing the benefits. A typical rotation might include a cycle of legumes, cereals, root vegetables and cover crops. The specific rotation will depend on the local climate, soil type and the farmer's goals. For instance, a three-year rotation might include corn (heavy feeder), followed by beans (nitrogen fixer) and then a cover crop like clover or rye to protect and enrich the soil during the off-season.

Integrating Seed Saving and Crop Rotation for Sustainability

The integration of seed saving and crop rotation practices creates a synergistic effect that enhances the sustainability of farming systems. When combined, these practices promote soil health, preserve genetic diversity and improve the resilience of crops, leading to more stable and productive agricultural systems over time.

Adaptive Seed Selection: seed saving allows farmers to select and propagate crop varieties that are well-suited to their specific crop rotation plans and local environmental conditions. By choosing seeds from plants that thrive under particular rotation sequences, farmers can develop crops that are better adapted to their soil, climate and pest pressures.

Rotation-Informed Seed Saving: crop rotation informs the selection of which seeds to save based on the performance of crops in different rotation sequences. For example, if a particular variety of wheat performs exceptionally well following a legume crop, saving seeds from that wheat variety can help ensure similar success in future rotations.

Enhanced Resilience: the combination of seed saving and crop rotation enhances the resilience of farming systems. Diverse crop rotations reduce the risk of crop failure due to pests, diseases or environmental stresses, while seed saving ensures that farmers have access to a diverse pool of genetic resources to respond to changing conditions.

Seed saving and crop rotation are essential practices for sustainable, long-term food production. These techniques not only promote soil health and reduce the need for external inputs but also preserve the genetic diversity of crops and enhance their resilience to pests, diseases and environmental changes. By integrating seed saving and crop rotation into their farming practices, farmers can build more sustainable, productive and resilient agricultural systems that are capable of supporting human needs for generations to come.

Long-term food production is about more than just growing crops; it's about building a sustainable, resilient system that can provide for your needs year after year. By focusing on the key aspects of sustainability, biodiversity and resilience and by incorporating advanced techniques such as agroforestry, perennial systems, seed saving and crop rotation, you can create a food production system that not only meets your immediate needs but also ensures the long-term health and productivity of your land.

These practices help reduce your reliance on external inputs, improve the ecological balance of your farm and enhance your overall food security. As you implement these techniques, remember that long-term food production is a journey, not a destination. Continuous learning, adaptation and improvement are key to ensuring that your food production system remains viable and productive for generations to come.

Exercise Chapter 12
Planning a Year-Round Crop Rotation

Objective: learn how to design and implement a year-round crop rotation plan that maximizes soil fertility, minimizes pest and disease issues and enhances overall farm productivity.

Materials Needed: notebook or digital planner, garden map or layout, list of crops to be planted, calendar and reference materials on crop families and their specific nutrient needs.

1. Assess Your Garden Layout: begin by drawing a map of your garden or farm, noting the dimensions, current soil conditions and any existing structures. Identify which areas receive the most sunlight, which are more shaded and where water tends to collect.

2. Choose Your Crops: make a list of the crops you plan to grow throughout the year, considering your local climate and growing season. Categorize these crops by their families (e.g., legumes, brassicas, nightshades) and their specific nutrient requirements (e.g., nitrogen-heavy, root crops).

3. Design the Crop Rotation Plan: with your map and crop list in hand, plan where each crop will be planted throughout the year. Rotate crops based on their families to prevent soil depletion and reduce pest and disease buildup. For example, follow nitrogen-fixing legumes with heavy feeders like corn and plant root crops in areas where leafy greens were previously grown.

4. Schedule Planting and Harvesting: create a calendar that outlines when each crop will be planted and harvested. Consider the timing of crop maturation and how quickly you can replant after harvesting. Plan for successive planting where possible to ensure continuous production throughout the year.

5. Document and Adjust: write down your crop rotation plan in detail, including the rationale behind each decision. Throughout the growing season, monitor the progress of your crops and make adjustments as needed based on their performance and any unforeseen challenges.

Deliverable: prepare a concise report of your year-round crop rotation plan. Include your garden map, crop list with locations and the planting/harvesting schedule. Reflect on anticipated challenges and how you'll address them. Note any adjustments made during the season and their outcomes.

This exercise will help you develop a comprehensive crop rotation plan tailored to your specific garden, enhancing soil health and crop productivity year after year.

Chapter 13
Personal Care and Hygiene

Maintaining personal care and hygiene is not just a matter of comfort, it is crucial for your health, well-being and survival, especially in challenging environments. In situations where access to modern conveniences may be limited, understanding how to manage personal hygiene becomes even more essential. Proper hygiene practices help prevent the spread of diseases and maintain overall health, ensuring that you can stay focused and effective in survival situations.

This chapter will explore the importance of personal care, delve into the basics of making your own hygiene products like soaps and detergents and cover effective waste management strategies that help maintain a clean and healthy environment.

Introduction to Personal Care

Personal care and hygiene are foundational to maintaining health and preventing illness, particularly in environments where resources are scarce. Good hygiene reduces the risk of infections and diseases, which can be critical in survival situations. Understanding the tools and materials needed, along with basic hygiene practices, can significantly enhance your ability to stay healthy.

Cultural practices and environmental conditions significantly influence personal hygiene routines. For instance, in water-scarce areas, traditional methods of water conservation, such as using minimal water for washing or employing techniques like 'dry bathing' with a cloth, become crucial. Additionally, cultural norms may dictate specific hygiene practices, such as the use of natural materials for cleaning or the importance of ritual cleansing. Understanding and respecting these practices can enhance the effectiveness of your hygiene routines, particularly when adapting to new environments or when resources are limited.

This section will guide you through the essential aspects of personal care, focusing on practical, everyday routines that ensure cleanliness and well-being, regardless of the circumstances.

Importance of Hygiene

Hygiene is the first line of defense against a wide range of diseases and infections. Proper hygiene practices help to prevent the spread of pathogens, which can cause illnesses ranging from minor skin infections to severe diseases like cholera or hepatitis. Maintaining good hygiene reduces the risk of cross-contamination and is particularly critical in environments where medical resources may be limited.

The impact of hygiene extends beyond individual health to the well-being of entire communities. In close-knit environments, such as family units, camps or survival groups, poor hygiene practices by even one person can quickly lead to outbreaks of disease, affecting everyone in the group. For this reason, educating individuals on the importance of personal hygiene is vital for communal health.

Moreover, personal hygiene has psychological benefits. In stressful situations, such as during a crisis or survival scenario, maintaining a routine that includes regular hygiene practices can provide a sense of normalcy and control. This, in turn, can reduce anxiety, boost morale and help individuals cope better with the challenges they face.

Tools and Materials for Personal Care

Effective personal care requires the right tools and materials, many of which are commonplace but can become invaluable in a survival situation. Here's a breakdown of essential items:

Soap and Detergents: these are the most basic yet crucial components of hygiene. Soaps help remove dirt, oils and microbes from the skin, while detergents are used for cleaning clothes and other materials. Understanding the properties of different types of soaps and detergents, as well as how to make them from natural ingredients, can be life-saving when commercial products are unavailable.

Water: clean water is essential for personal hygiene. It is used not only for drinking but also for washing hands, bathing, cleaning wounds and laundering clothes. In situations where water is scarce, knowing how to purify and conserve water is crucial.

Toothbrush and Toothpaste: oral hygiene is often overlooked in survival scenarios, but it is vital for preventing dental problems, which can become serious if left untreated. In the absence of commercial products, alternatives such as baking soda or natural chewing sticks can be used to maintain oral hygiene.

Nail Clippers and Scissors: keeping nails trimmed and clean prevents the accumulation of dirt and bacteria that can cause infections. Similarly, scissors can be used for trimming hair, managing wounds or cutting bandages, making them indispensable in any personal care kit.

Sanitary Supplies: for women, menstrual hygiene management is crucial. Reusable cloth pads, menstrual cups or even DIY solutions must be considered when access to commercial sanitary products is limited.

In addition to conventional tools, there are several alternative or improvised hygiene solutions that can be invaluable in extreme situations. For example, natural toothbrush substitutes like chew sticks, commonly made from twigs of certain trees, can effectively clean teeth when toothbrushes are unavailable. Additionally, in the absence of soap, ash or clay can be used as a natural cleaning agent. Ash has mild abrasive properties and can help remove dirt and oils from the skin, while clay can be used as a natural scrub, absorbing oils and impurities.

Basic Hygiene Practices

The foundation of personal care lies in basic hygiene practices, which should be adhered to consistently, regardless of the circumstances. These include:

Handwashing: regular handwashing with soap and water is one of the most effective ways to prevent the spread of diseases. In situations where soap is not available, using ash or hand sanitizers can be an effective alternative. It is important to wash hands before eating, after using the toilet and after any activity that exposes the hands to potential contaminants.

Bathing: regular bathing helps to remove dirt, sweat and oils from the skin, preventing skin infections and rashes. In survival scenarios where water is scarce, sponge baths or wiping down with a damp cloth can help maintain cleanliness.

Oral Hygiene: brushing teeth at least twice a day is essential for preventing tooth decay and gum disease. When toothbrushes are not available, alternatives such as using a clean cloth or chewing sticks can be employed.

Nail Care: keeping nails short and clean reduces the risk of harboring dirt and bacteria. It also prevents accidental scratches, which can lead to infections.

Wound Care: proper care of cuts, scrapes and other injuries is vital to prevent infections. This includes cleaning wounds with clean water, applying antiseptic if available and covering them with a clean bandage.

Concluding, personal care and hygiene are foundational to health and well-being, particularly in challenging situations where access to modern conveniences may be limited. By understanding the importance of hygiene, utilizing the appropriate tools and materials and adhering to basic hygiene practices, individuals can maintain their health and significantly reduce the risk of disease. These practices are not only essential for individual survival but also for the health of the community, making personal care a critical component of any preparedness plan.

Making Soaps and Detergents

The art of making soaps and detergents is both a practical skill and a creative endeavor that has been practiced for centuries. These products are essential for maintaining hygiene and cleanliness and being able to produce them independently can be invaluable, especially in situations where commercial products are unavailable or when aiming for a more self-sufficient lifestyle. This section will guide you through the materials needed, the production procedures and the best practices for using and storing homemade soaps and detergents.

Required Materials

To make soaps and detergents, a few key ingredients are necessary, each playing a specific role in the final product's effectiveness and safety. Here's an overview of the primary materials:

- **Fats and Oils**: the foundational ingredient in soap making, fats and oils (such as olive oil, coconut oil or lard) react with lye to create soap through a process called saponification. Different oils contribute different properties to the soap; for instance, coconut oil adds hardness and lather, while olive oil is known for its moisturizing qualities.
- **Lye (Sodium Hydroxide for Soap)**: lye is a crucial ingredient in traditional soap making. When lye is mixed with fats or oils, it triggers the chemical reaction of saponification, which results in soap. It's important to handle lye with care, as it is a caustic substance that can cause burns if mishandled.
- **Potassium Hydroxide**: while sodium hydroxide (lye) is used for making solid soaps, potassium hydroxide is often used in the production of liquid soaps. The choice between the two depends on whether a solid or liquid end product is desired.
- **Water**: water is used to dissolve the lye before it is mixed with oils. The quality of water (distilled is preferred) can affect the final product, as impurities in tap water can interfere with the saponification process.
- **Essential Oils and Fragrances**: these are added to soaps to impart a pleasant scent. Essential oils such as lavender, tea tree or eucalyptus also add natural antimicrobial properties to the soap.
- **Colorants and Additives**: natural colorants like clays, activated charcoal or botanical extracts can be used to color soaps. Additives like oatmeal, aloe vera or honey can enhance the soap's benefits for the skin.
- **Detergents**: detergents differ from traditional soap in that they are synthetic cleaners. The main ingredient in detergents is a surfactant, which is responsible for breaking down grease and dirt. Common surfactants include sodium lauryl sulfate (SLS) and sodium laureth sulfate (SLES). Detergents may also include builders like phosphates, which enhance cleaning power by softening water.

Production Procedure

Making soap and detergent at home is a straightforward process that requires attention to detail and safety, particularly when handling caustic substances like lye. Here's a step-by-step guide:

Soap Making:
1. **Preparation**: start by gathering all necessary materials and tools. Wear protective gear, including gloves and goggles, to protect against lye splashes. Ensure your workspace is well-ventilated.
2. **Mixing the Lye Solution**: carefully measure out the lye and water separately. Slowly add the lye to the water (never the reverse) while stirring until fully dissolved. The mixture will heat up quickly and release fumes, so be cautious and work in a well-ventilated area.
3. **Heating the Oils**: while the lye solution cools, measure and heat your chosen oils or fats in a large pot. Heat the oils to a specific temperature range, typically between 100°F and 110°F, which should match the temperature of the lye solution.
4. **Combining Lye and Oils**: once both the lye solution and oils are at the correct temperature, slowly pour the lye solution into the oils. Stir continuously until the mixture reaches "trace," where it thickens to the consistency of pudding. This indicates that saponification is well underway.
5. **Adding Fragrances and Additives**: at trace, you can add essential oils, colorants or other additives. Stir thoroughly to ensure even distribution throughout the soap mixture.
6. **Molding**: pour the soap mixture into molds, smoothing the top if necessary. Cover the molds and insulate them to allow the soap to cool slowly and harden over the next 24-48 hours.
7. **Curing**: once the soap has hardened, remove it from the molds and cut it into bars if necessary. Allow the bars to cure in a cool, dry place for 4-6 weeks. This curing time allows the soap to fully saponify and dry out, improving its hardness and longevity.

Detergent Making:
1. **Mixing Ingredients**: combine the surfactant with water in a large container, stirring until fully dissolved. If using a powdered detergent formula, mix the powders thoroughly in a large bowl.
2. **Adding Builders and Fragrances**: if desired, add builders like washing soda or borax to enhance cleaning power. Fragrances or essential oils can also be added for a pleasant scent.
3. **Storing**: pour the liquid detergent into bottles or store the powdered detergent in an airtight container. Label your containers with the ingredients and usage instructions.

Use and Storage

Once your homemade soaps and detergents are ready, proper storage and use are key to maintaining their effectiveness and longevity:

- **Storage**: store soaps in a cool, dry place. To extend the life of your soap, keep it on a well-drained soap dish between uses. Detergents should be stored in sealed containers to prevent moisture absorption, which can cause clumping in powdered detergents or degradation in liquid formulas.

- **Usage**: homemade soaps can be used just like commercial soaps but with the added benefit of knowing exactly what ingredients are in them. Use detergents according to the type of cleaning – whether for laundry, dishes or general household cleaning – and adjust the amount based on the soil level and water hardness.

Making your own soaps and detergents not only provides a reliable source of hygiene products but also allows for customization to suit your skin type, scent preferences and cleaning needs. It's a rewarding skill that contributes to a more self-sufficient lifestyle and can even lead to cost savings over time.

Waste Management

Effective waste management is a critical component of maintaining hygiene, especially in self-sufficient or survival scenarios. Properly handling waste prevents the spread of diseases, protects the environment and conserves valuable resources. This section will explore various waste management techniques, including disposal methods, composting, recycling and ensuring safety and hygiene throughout the process.

Disposal Techniques

The first step in managing waste is understanding the different types of waste and the appropriate methods for disposing of each. Waste can generally be categorized into organic waste, inorganic waste and hazardous waste, each requiring different disposal techniques:

- **Organic Waste**: this includes food scraps, plant materials and other biodegradable items. Organic waste is best managed through composting, which allows it to decompose naturally, turning it into valuable compost that can be used to enrich the soil.
- **Inorganic Waste**: this category includes materials like plastic, metal, glass and other non-biodegradable items. Proper disposal of inorganic waste often involves recycling or reusing materials to minimize environmental impact. In survival situations, finding ways to repurpose inorganic materials can be particularly valuable.
- **Hazardous Waste**: this includes substances like batteries, chemicals and certain medical waste. Hazardous waste requires careful handling and disposal to prevent contamination of the environment and to protect human health. In remote or survival settings, hazardous waste should be securely stored until it can be safely disposed of, typically by taking it to a designated facility when possible.

Composting and Recycling

Composting and recycling are two essential waste management practices that not only reduce the amount of waste that ends up in landfills but also turn waste into valuable resources.

Composting:

- **Process**: composting involves collecting organic waste and allowing it to decompose naturally over time. This process is facilitated by microorganisms that break down the material into nutrient-rich compost, which can be used to improve soil health.
- **Materials**: the composting process works best with a balanced mix of "green" materials (like vegetable scraps, coffee grounds and fresh grass clippings) and "brown" materials (like dry leaves, straw and cardboard). The green materials provide nitrogen, while the brown materials add carbon; both are necessary for effective decomposition.
- **Benefits**: composting reduces the volume of waste, decreases the need for chemical fertilizers and improves soil structure, water retention and aeration. It's a sustainable practice that contributes to a closed-loop system in which waste is reused and recycled within the environment.

Recycling:
- **Sorting**: effective recycling begins with sorting waste materials into categories such as plastics, metals, glass and paper. In some cases, further sorting by type (like separating different types of plastics) may be necessary to ensure the material can be properly processed.
- **Repurposing**: in a self-sufficient setting, finding ways to repurpose materials can be crucial. For example, glass jars can be reused for storage, metal cans can be fashioned into tools or containers and plastic bottles can be converted into plant pots or even building materials.
- **Environmental Impact**: recycling reduces the need for raw materials, conserves energy and minimizes pollution. It's a key component of sustainable living and can significantly lower the environmental footprint of a homestead or survival situation.

Safety and Hygiene

Maintaining safety and hygiene during waste management is crucial to prevent contamination and the spread of diseases:

- **Handling Waste**: always wear protective gloves when handling waste, especially hazardous materials. Ensure that waste storage areas are secure and out of reach of children and animals. Regularly disinfect areas where waste is stored to prevent the buildup of harmful bacteria or pests.
- **Sanitation**: for waste that cannot be immediately composted or recycled, ensure that it is stored in a sealed container to minimize odors and prevent attracting pests. Human waste, in particular, requires careful management, constructing a simple latrine or using composting toilets can provide a sanitary solution in remote or emergency settings.
- **Pest Control**: unmanaged waste can attract pests like rodents, insects and larger animals, which can spread disease. Regularly remove waste from living areas and use traps or natural deterrents to keep pests at bay. Compost piles should be monitored to ensure they are not attracting unwanted wildlife.
- **Water Safety**: waste management also includes ensuring that water sources are protected from contamination. Avoid disposing of waste near streams, rivers or wells and ensure that any waste disposal systems are located downhill from water sources to prevent runoff contamination.

Effective waste management is a cornerstone of personal care and hygiene, especially in survival or self-sufficient living scenarios. By implementing proper disposal techniques, embracing composting and recycling and maintaining strict safety and hygiene standards, you can manage waste effectively while protecting your health and the environment. These practices not only contribute to a cleaner, more sustainable living environment but also enhance the overall resilience and self-sufficiency of your lifestyle.

In conclusion, personal care and hygiene are foundational aspects of maintaining health and well-being, especially in situations where modern conveniences are not available. By understanding the importance of hygiene, learning to make your own soaps and detergents and implementing effective waste management practices, you can ensure a clean and healthy environment. These skills not only protect your health but also contribute to a sustainable and resilient lifestyle, enabling you to thrive even in challenging conditions.

As you move forward, remember that personal care and hygiene are ongoing practices that require regular attention and refinement. Make it a habit to practice these skills routinely, even in non-emergency situations, to ensure that you are fully prepared when it matters most. By staying vigilant about hygiene and continuously improving your techniques, you can maintain your health, enhance your resilience and ensure that you are always ready to face whatever challenges arise.

Exercise Chapter 13
Making Your Own Natural Soap

Objective: learn how to create natural soap using basic ingredients and traditional methods.

Materials Needed: fats or oils (like olive oil, coconut oil or lard), sodium hydroxide (lye), distilled water, essential oils for fragrance (optional), natural colorants (like activated charcoal or turmeric), soap molds, protective gear (gloves and goggles), a digital scale, a thermometer, mixing bowls and an immersion blender or sturdy spoon.

1. Prepare the Workspace: set up a clean, well-ventilated workspace. Gather all necessary materials and put on protective gear, including gloves and goggles, to protect against lye, which is caustic.

2. Mix the Lye Solution: carefully measure distilled water into a heat-resistant container. Gradually add lye to the water, stirring continuously. Ensure good ventilation during this process as the solution will heat up and release fumes. Let the lye solution cool to the required temperature.

3. Heat the Oils: measure the chosen oils and heat them in a large pot. The temperature of the oils should match that of the lye solution, typically between 100°F and 110°F.

4. Combine the Ingredients: slowly pour the lye solution into the oils. Stir or use an immersion blender until the mixture reaches "trace," where it thickens to a pudding-like consistency.

5. Add Fragrances and Colorants: stir in any essential oils or natural colorants you wish to include, ensuring they are evenly distributed.

6. Pour into Molds: pour the mixture into soap molds, smoothing the top if needed. Insulate the molds with a towel and let the soap harden for 24-48 hours.

7. Cure the Soap: after removing the soap from the molds, cut it into bars if necessary. Allow the bars to cure in a cool, dry place for 4-6 weeks, during which time they will harden and become ready for use.

Deliverable: write a brief report on your soap-making experience. Note any challenges and how you resolved them. Include photos of your process and final soap. Reflect on the outcome and consider any improvements.

This exercise will help you gain confidence in making natural soap, providing you with a valuable skill for self-sufficiency and personal care.

Chapter 14
Natural Medicines and First Aid

Natural medicine has been the foundation of healthcare for millennia, long before the advent of modern pharmaceuticals. In many parts of the world, it remains the primary means of treating common ailments and maintaining health. The use of plants, herbs and other natural substances offers an accessible, sustainable way to manage health, especially when professional medical help is out of reach.
Understanding and utilizing natural remedies can empower you to take charge of your health and well-being, using the resources available in your environment.

Importance of Natural Remedies

Natural remedies are often more accessible, sustainable and cost-effective than synthetic medicines. They can be cultivated or foraged locally, reducing dependence on commercial pharmaceuticals and ensuring that essential medicines are available when needed most. Additionally, natural remedies often have fewer side effects and are better tolerated by the body, making them a preferred option for long-term health maintenance.

Moreover, understanding natural remedies enhances self-reliance, a key aspect of survival and preparedness. In scenarios where access to healthcare facilities is compromised, knowing how to treat minor injuries, illnesses or even chronic conditions with natural remedies can be life-saving. This knowledge not only provides immediate solutions but also contributes to a holistic approach to health, focusing on prevention and the natural strengthening of the body's defenses.

Common Medicinal Plants

The world of medicinal plants is vast, with countless species offering various therapeutic benefits. Here are a few essential medicinal plants that are both effective and commonly found in many regions:

- **Aloe Vera**: known for its soothing properties, aloe vera is widely used to treat burns, cuts and skin irritations. The gel inside its leaves can be applied directly to the skin to promote healing and reduce inflammation.
- **Garlic**: garlic is a powerful natural antibiotic and antifungal agent. It can be used to boost the immune system, lower blood pressure and treat infections. Consuming raw garlic or using garlic-infused oil can provide these benefits.
- **Echinacea**: this plant is often used to strengthen the immune system and fight off colds and respiratory infections. Echinacea can be consumed as a tea, tincture or capsule.
- **Calendula**: the bright orange flowers of calendula are not only beautiful but also medicinal. Calendula is excellent for treating cuts, burns and rashes due to its anti-inflammatory and antimicrobial properties. It is often used in salves and creams.
- **Peppermint**: peppermint is commonly used to soothe digestive issues, reduce nausea and relieve headaches. It can be consumed as a tea or applied topically as an essential oil to relieve tension headaches.
- **Lavender**: known for its calming effects, lavender is used to reduce stress, anxiety and insomnia. It is also useful in treating minor burns and insect bites. Lavender can be used in teas, as an essential oil or in dried form for aromatherapy.

Harvesting and Preparation

Proper harvesting and preparation are crucial for maximizing the medicinal properties of plants. The time of year, the part of the plant used and the method of preparation can all influence the effectiveness of a remedy.

Harvesting: the best time to harvest medicinal plants often depends on the specific plant and the part being used. For example, leaves are typically harvested just before the plant flowers, while roots are best harvested in the fall when the plant's energy is concentrated below ground. Always use clean, sharp tools to avoid damaging the plant and only take what you need to allow the plant to continue growing.

Drying: many medicinal plants need to be dried before use, especially if they are to be stored for later use. Drying should be done in a well-ventilated area, away from direct sunlight, to preserve the plant's active compounds. Once dried, store the plants in airtight containers, away from light and moisture.

Infusions and Decoctions: infusions are made by steeping plant material, typically leaves or flowers, in hot water, similar to making tea. Decoctions are used for tougher plant materials like roots or bark and involve simmering the material in water for an extended period. Both methods extract the active compounds from the plant, making them easier to consume.

Tinctures: tinctures are concentrated herbal extracts made by soaking herbs in alcohol or vinegar for several weeks. The resulting liquid contains the plant's active ingredients and can be used in small doses for medicinal purposes.

Salves and Balms: salves and balms are made by infusing medicinal plants into oils and then thickening the mixture with beeswax or another thickening agent. These are used topically to treat skin conditions, wounds or muscle pain.

Building a First Aid Kit

A well-prepared first aid kit is a cornerstone of effective emergency response. It is not just about having a box of supplies; it's about ensuring you have the right tools at your disposal to handle a wide range of medical situations, from minor cuts to life-threatening emergencies. The importance of a comprehensive first aid kit cannot be overstated, particularly in situations where professional medical help might be delayed or unavailable.

Essential Elements

A comprehensive first aid kit should be equipped to handle various types of injuries and medical emergencies. Here's a breakdown of the essential elements every first aid kit should include:

Wound Care Supplies:
- **Adhesive Bandages**: a variety of sizes to cover small cuts, blisters and abrasions.
- **Sterile Gauze Pads**: these are crucial for dressing larger wounds and controlling bleeding.
- **Elastic Bandages**: useful for securing dressings and providing support for sprains or strains.
- **Adhesive Tape**: for securing bandages or gauze pads in place.
- **Antiseptic Wipes**: to clean wounds and reduce the risk of infection.
- **Antibiotic Ointment**: helps to prevent infection in minor cuts, scrapes and burns.

Medications:
- **Pain Relievers**: over-the-counter medications such as ibuprofen or acetaminophen for pain relief.
- **Antihistamines**: to treat allergic reactions and reduce symptoms like itching or swelling.
- **Hydrocortisone Cream**: for treating insect bites, rashes or skin irritations.
- **Aloe Vera Gel**: useful for soothing burns and sunburns.
- **Any Prescription Medications**: ensure you have an adequate supply of any medications you or your family members regularly take.

Tools and Instruments:

- **Scissors**: for cutting tape, gauze or clothing if needed.
- **Tweezers**: essential for removing splinters, ticks or other foreign objects from the skin.
- **Thermometer**: a digital thermometer is crucial for monitoring fevers.
- **Safety Pins**: for securing bandages or slings.
- **CPR Mask**: to provide safe and effective mouth-to-mouth resuscitation.

Specialized Items:
- **Tourniquet**: for controlling severe bleeding in an emergency.
- **Instant Cold Packs**: to reduce swelling and pain from injuries like sprains or insect bites.
- **Burn Dressings**: special dressings to treat burns and protect the injured area from infection.
- **Finger Splints**: for immobilizing injured fingers to prevent further damage.
- **Emergency Blanket**: helps to prevent hypothermia by retaining body heat in cold conditions.

Kit Organization

The effectiveness of a first aid kit is greatly enhanced by how well it is organized. In an emergency, you need to be able to find and access the right supplies quickly. Here are some tips for organizing your kit:

- **Categorize by Injury Type**: organize your kit into sections, such as wound care, medications and tools. This way, you can quickly find the supplies you need for a specific type of injury.
- **Label Everything**: clearly label each section and the contents within. This can be particularly helpful in stressful situations where you may need to find items quickly.
- **Use Transparent Pouches**: consider storing items in clear, resealable bags or pouches within the kit. This keeps everything visible and easily accessible.
- **Regularly Check and Restock**: make it a habit to regularly check your first aid kit for expired medications or used items and restock as necessary. A well-maintained kit is always ready for use.

Emergency Use

Having a fully stocked first aid kit is only part of the equation; knowing how to use its contents effectively is just as crucial. Here are some steps to ensure you're prepared to use your kit in an emergency:

- **Familiarize Yourself with the Kit**: spend time learning where each item is located within the kit and understand its purpose. This familiarity will save valuable time during an emergency.
- **Take a First Aid Course**: consider enrolling in a first aid and CPR course to gain hands-on experience. These courses often cover the basics of wound care, CPR and how to handle common medical emergencies.
- **Practice Scenarios**: regularly practice emergency scenarios where you use your first aid kit. This can help build confidence and ensure that you can perform under pressure.

- **Instruct Others**: make sure that other family members or group members are also familiar with the kit and know how to use it. In an emergency, you may need to rely on someone else to assist with first aid.

Building and maintaining a comprehensive first aid kit is an essential step in ensuring that you are prepared for emergencies. By organizing your kit effectively and familiarizing yourself with its contents, you can respond quickly and efficiently to a wide range of medical situations. This preparation not only helps to mitigate the impact of injuries and illnesses but also provides peace of mind, knowing that you are ready to handle whatever comes your way.

First Aid Techniques

First aid techniques are critical skills that can make a significant difference in emergency situations, often bridging the gap between the onset of an injury or illness and professional medical treatment. Whether you're dealing with minor injuries or life-threatening emergencies, understanding and effectively applying first aid techniques can save lives and prevent complications. This section delves into some of the most essential first aid techniques, including treating wounds and bleeding, managing fractures and immobilizations and addressing shock and hypoglycemia.

Treating Wounds and Bleeding

Cleaning and Dressing Wounds: proper wound care is essential to prevent infection and promote healing. Here's how to effectively manage wounds:
1. **Initial Cleaning**: before dressing a wound, it's crucial to clean it thoroughly. Use clean water or a saline solution to gently rinse away dirt, debris or foreign objects. Avoid using strong antiseptics like hydrogen peroxide directly on the wound, as they can damage tissue and delay healing.
2. **Disinfection**: after cleaning, apply an antiseptic, such as iodine or alcohol-free antiseptic wipes, around the wound area to kill any remaining bacteria. Be careful not to apply harsh chemicals directly into the wound itself.
3. **Dressing the Wound**: choose an appropriate dressing based on the wound's size and location. For minor cuts, adhesive bandages are sufficient. For larger wounds, use sterile gauze pads and secure them with medical tape. Ensure the dressing is snug but not too tight, allowing for circulation.
4. **Monitoring**: keep an eye on the wound for signs of infection, such as redness, swelling, increased pain or pus. Change the dressing regularly and ensure the wound stays clean and dry.

Controlling Bleeding: in cases of severe bleeding, quick action is required to prevent significant blood loss. Here's what to do:
1. **Direct Pressure**: the first and most effective method to control bleeding is applying direct pressure to the wound using a clean cloth or sterile gauze. Hold the pressure firmly until the bleeding stops. If the cloth becomes soaked with blood, add another layer rather than removing the initial one, to maintain pressure.

2. **Elevation**: if possible, elevate the injured limb above the level of the heart to reduce blood flow to the area, which can help slow the bleeding.
3. **Tourniquet Use**: if direct pressure and elevation are not effective and the bleeding is life-threatening, applying a tourniquet may be necessary. Place the tourniquet above the wound, close to the body and tighten it until the bleeding stops. Note the time when the tourniquet was applied, as it should be loosened every 20 minutes to avoid tissue damage.
4. **Aftercare**: once the bleeding is controlled, keep the person warm and seek immediate medical assistance. Continue to monitor for shock, which can occur as a result of significant blood loss.

Fractures and Immobilizations

Fractures or broken bones, require careful management to prevent further injury and reduce pain. Proper immobilization of the affected area is crucial.

Identifying a Fracture:
1. **Signs and Symptoms:** common signs of a fracture include intense pain, swelling, deformity and an inability to move the affected limb. In some cases, the bone may protrude through the skin, known as an open or compound fracture.
2. **Assessing the Injury**: before attempting to move the injured person, assess the severity of the fracture. If the fracture is open, treat any bleeding first by applying a sterile dressing around the wound, avoiding direct pressure on the bone.

Immobilizing the Fracture:
1. **Splinting**: a splint is used to immobilize the fractured area to prevent further injury and reduce pain. If a professional splint is not available, you can improvise using rigid materials like sticks, boards or rolled-up newspapers. Pad the splint to prevent it from pressing directly on the skin. Secure the splint with cloth strips, bandages or belts, tying it above and below the fracture site. Ensure the splint is snug but not too tight, as this can cut off circulation.
2. **Slings for Arm Fractures**: for fractures of the arm or shoulder, creating a sling with a triangular bandage can help immobilize the limb and reduce pain. Position the sling so that the arm is bent at a 90-degree angle across the chest, with the hand slightly elevated.
3. **Avoiding Movement**: it's crucial to minimize movement of the fractured area. Encourage the injured person to stay still and avoid moving them unless absolutely necessary, such as in a life-threatening situation where evacuation is required.

Shock and Hypoglycemia Treatment

Shock is a life-threatening condition that can occur after a severe injury, blood loss or as a result of a serious medical condition. Hypoglycemia or low blood sugar, is another critical condition that requires immediate attention.

Treating Shock:
1. **Recognizing Shock**: symptoms of shock include rapid breathing, pale or clammy skin, weakness, dizziness, confusion and a weak but rapid pulse. Shock can progress quickly, so it's vital to act fast.
2. **Positioning**: lay the person down on their back and elevate their legs about 12 inches to help improve blood flow to vital organs. If they have a head, neck or spinal injury, avoid moving them and keep them flat.
3. **Maintaining Warmth**: cover the person with a blanket or clothing to maintain body temperature, but avoid overheating. This helps prevent hypothermia, which can exacerbate shock.
4. **Reassurance**: Keep the person calm and reassured. Anxiety can worsen shock, so speak softly and clearly, explaining what you're doing to help them.
5. **Emergency Response**: if the person is unconscious or shows signs of severe shock, call for emergency medical help immediately. Continue to monitor their breathing and pulse and be prepared to administer CPR if they stop breathing.

Managing Hypoglycemia:
1. **Identifying Hypoglycemia**: symptoms of hypoglycemia include sweating, trembling, weakness, confusion, irritability and, in severe cases, loss of consciousness. It often occurs in individuals with diabetes but can affect anyone.
2. **Quick Sugar Source**: if the person is conscious and able to swallow, give them a quick source of sugar, such as glucose tablets, fruit juice, candy or sugar dissolved in water. This helps raise their blood sugar level rapidly.
3. **Monitoring**: after administering sugar, monitor the person's condition closely. If symptoms persist or worsen, seek medical assistance. For individuals with diabetes, follow their specific management plan.
4. **Preventing Hypoglycemia**: encourage regular meals and snacks, particularly if the person is engaged in physical activity or if they are diabetic. Always have a source of quick sugar available in case of an emergency.

First aid techniques are essential skills that everyone should know. Whether dealing with minor injuries or life-threatening emergencies, being able to apply these techniques effectively can save lives and prevent further harm. By understanding how to treat wounds and control bleeding, immobilize fractures and manage conditions like shock and hypoglycemia, you can be better prepared to handle medical emergencies confidently. Regular practice and staying informed about first aid procedures are key to ensuring that you are always ready to act when needed.

In conclusion, the knowledge of natural medicines and first aid is an essential part of self-sufficiency and preparedness. Whether dealing with minor injuries, serious emergencies or simply maintaining health in a challenging environment, these skills empower you to take control of your well-being and that of those around you. Continuously practicing and refining these skills ensures that you are always ready to respond effectively in any situation.

Exercise Chapter 14
Creating a Herbal First Aid Kit

Objective: assemble a practical herbal first aid kit to treat common ailments and injuries using natural remedies.

Materials Needed: small jars or tins with lids, dropper bottles, labels, marker, beeswax, carrier oil (like olive or coconut oil), essential oils, dried herbs (such as calendula, yarrow, lavender, comfrey), cheesecloth and a small bag or box.

1. Research and select herbs: start by researching the medicinal properties of various herbs. Choose a few that are well-suited to your needs, such as calendula for healing skin wounds and rashes, yarrow for stopping bleeding and treating cuts, lavender for calming and antiseptic properties and comfrey for supporting the healing of bones and tissues.

2. Prepare herbal remedies: using your selected herbs, create different remedies. For a herbal oil, infuse dried herbs in a carrier oil for several weeks, strain and store in a bottle. For a salve, mix your infused oil with melted beeswax to create a salve; pour into tins and allow to cool and solidify. For a tincture, steep herbs in alcohol for a few weeks, strain and store in dropper bottles; use tinctures internally or topically depending on the herb.

3. Assemble the kit: organize your herbal preparations into the small containers and label each one clearly with the contents and usage instructions. Include salves for skin issues, tinctures for internal and topical use, herbal oils for massages or skin application, essential oils for their various properties and basic first aid supplies like bandages, gauze, scissors and tweezers.

4. Store and maintain: keep your herbal first aid kit in a cool, dry place. Regularly check the contents to ensure everything is still fresh and effective, replacing items as needed.

Deliverable: write a brief report on your experience creating the herbal first aid kit. Note any challenges and how you overcame them. Include photos of your process and the finished kit. Reflect on how well the kit meets your needs and any changes you might make in the future.

Creating a herbal first aid kit not only equips you with natural remedies but also deepens your understanding of the healing properties of plants. Practice using these remedies so you're prepared to confidently address health needs as they arise.

Chapter 15
Natural Resource Management

In the quest for self-sufficiency and sustainable living, managing natural resources is a crucial skill that extends far beyond simple survival. It encompasses the knowledge and practices necessary to use, conserve and renew the natural assets available in your environment. Whether you're managing a forest, interacting with wildlife or conserving water, the principles of natural resource management are fundamental to ensuring that these resources are available not just for your immediate needs, but for future generations as well. Effective natural resource management begins with an understanding of what resources are available and how they can be used sustainably. This includes everything from wood for fuel and building, to wildlife as a food source and even the water and soil that support all life. By adopting sustainable practices, you can create a balanced system that meets your needs while maintaining the health of the environment.

This chapter will explore the principles of natural resource management, with a focus on forest and wildlife management. We'll delve into identifying resources, sustainable harvesting and conservation techniques that will help you make the most of your environment without depleting it. By the end of this chapter, you'll have a comprehensive understanding of how to manage natural resources effectively, ensuring that they remain abundant and healthy for years to come.

Introduction to Resource Management

Natural resource management is an essential aspect of sustainable living and survival, focusing on the careful and responsible use of natural resources to ensure they are available for future generations. This practice is deeply rooted in the understanding that our natural environment provides us with all the essentials for life – water, food, shelter and energy – and that these resources must be managed wisely to maintain ecological balance and support human well-being. In a world facing increasing environmental challenges such as climate change, deforestation and biodiversity loss, the principles of resource management are more important than ever.

Sustainable management of resources is the foundation of a resilient and enduring lifestyle, whether you are living off the grid, running a self-sufficient homestead or simply aiming to reduce your environmental footprint. It involves using resources at a rate at which they can naturally replenish, thereby avoiding the depletion of these critical assets. This concept is not only about conserving resources but also about enhancing the health of ecosystems to support long-term human survival.

One of the first steps in effective resource management is the identification of the resources available in your environment. This includes both natural and renewable resources such as water, soil, forests and wildlife, as well as non-renewable resources like minerals and fossil fuels. By recognizing what resources are present and understanding their abundance, quality and renewability, you can develop strategies to use them more efficiently. This identification process also involves assessing the current state of these resources, which may vary based on factors such as location, climate and existing land use practices. For instance, understanding the health of a forest ecosystem can inform decisions on sustainable logging practices, while assessing soil fertility can guide agricultural planning.

Once resources have been identified, the focus shifts to conservation techniques, which are critical for maintaining the health and availability of these resources over time. Conservation is about more than just limiting use; it's about actively managing and restoring ecosystems to ensure that they continue to function effectively. Techniques such as water conservation, soil management and habitat preservation are integral to this process.

Water conservation is particularly crucial in areas prone to drought or where water resources are limited. Practices such as rainwater harvesting, the use of efficient irrigation systems and the protection of natural water bodies are all important strategies. For example, implementing drip irrigation in agriculture can significantly reduce water use while maintaining crop yields. Additionally, protecting wetlands and forests that naturally filter and store water is vital for ensuring a stable and clean water supply.

Soil conservation is another key element of resource management. Healthy soil is the foundation of agriculture and forestry, supporting plant growth and maintaining ecological balance. Practices such as crop rotation, cover cropping, reduced tillage and the use of organic fertilizers help maintain soil fertility and prevent erosion. These techniques not only preserve the soil but also enhance its ability to sequester carbon, thus contributing to climate change mitigation.

Habitat preservation is essential for maintaining biodiversity, which in turn supports ecosystem services like pollination, pest control and nutrient cycling. Preserving natural habitats, such as forests, wetlands and grasslands, ensures that wildlife populations remain stable and that ecosystems continue to provide the resources and services humans rely on. In some cases, active restoration of degraded habitats may be necessary to bring back ecological functions that have been lost due to human activities.

Sustainable resource management also involves a deep understanding of the interconnectedness of different ecosystems and the cumulative impacts of human activities. For instance, deforestation not only depletes forest resources but also affects water cycles, soil health and biodiversity. Similarly, overfishing can lead to the collapse of marine ecosystems, impacting food security and livelihoods. Therefore, a holistic approach to resource management is required, one that considers the broader ecological, social and economic impacts of resource use.

In addition to these environmental considerations, resource management also encompasses the social and economic dimensions of sustainability. This means ensuring that resource use supports local communities, contributes to economic stability and does not exacerbate social inequalities. For example, community-based resource management, where local people are involved in decision-making processes, can lead to more equitable and effective management outcomes. It also fosters a sense of stewardship and responsibility for the environment, encouraging sustainable practices that benefit both people and the planet.

In conclusion, natural resource management is about creating a sustainable balance between human needs and the health of the environment. By understanding and applying the principles of resource management, you can contribute to the conservation of vital resources, support ecosystem resilience and ensure that these resources remain available for future generations. This approach not only enhances your own self-sufficiency but also plays a crucial role in addressing global environmental challenges, making it an essential component of any sustainable living or survival strategy.

Forest Management

Forest management is a crucial aspect of natural resource management, focusing on the sustainable use and conservation of forest ecosystems. Forests provide a wide range of resources and services, including timber, non-timber products, biodiversity conservation, water regulation, carbon sequestration and recreational opportunities. Effective forest management ensures that these resources are used in a way that maintains the health and vitality of the forest, supporting both ecological balance and human needs.

Identifying Tree Species

The first step in forest management is the identification of tree species within a given area. Understanding the composition of tree species is essential for several reasons: it informs decisions about which species to harvest, which to protect and how to manage the forest for multiple uses. Tree species vary widely in their ecological roles, economic value and response to management practices.

In many forests, certain tree species may be more valuable for timber production, while others may be critical for maintaining biodiversity or providing habitat for wildlife. For example, hardwood species such as oak, maple and walnut are often prized for their timber value, while conifers like pine and spruce may be more valuable for fast-growing pulpwood. Additionally, some species, such as willows and poplars, are important for stabilizing stream banks and preventing erosion.

Identifying tree species also involves understanding their growth patterns, reproductive strategies and ecological requirements. This knowledge helps forest managers make informed decisions about which species to promote or control, depending on the management objectives. For example, in a forest where biodiversity is a priority, managers might focus on protecting a diverse range of species, including those that are rare or threatened. In contrast, in a commercial forest, the focus might be on promoting species that are fast-growing and economically valuable.

Sustainable Cutting and Conservation

Sustainable cutting, also known as sustainable harvesting or logging, is a central practice in forest management. It involves the careful selection and removal of trees in a way that maintains the overall health of the forest and ensures that it can continue to provide resources and ecosystem services in the future. There are several approaches to sustainable cutting, each suited to different forest types and management goals:

- **Selective Cutting**: this method involves the removal of specific trees while leaving others intact. Selective cutting is often used to promote the growth of particular species, improve forest structure or reduce competition among trees. It is considered more sustainable than clear-cutting because it minimizes disruption to the forest ecosystem and allows for natural regeneration.
- **Clear-Cutting with Retention**: while traditional clear-cutting involves removing all trees in an area, clear-cutting with retention leaves behind certain trees or patches of forest to maintain habitat connectivity, protect water quality and promote natural regeneration. This approach is often used in managed forests where the goal is to produce timber while also conserving ecological values.
- **Shelterwood Cutting**: in shelterwood cutting, trees are removed in phases, allowing for a mix of young and old trees to coexist. This method promotes natural regeneration under the shelter of remaining trees, which can protect young seedlings from harsh environmental conditions and provide habitat for wildlife.
- **Coppicing**: this technique involves cutting trees down to their base, allowing them to regrow from the stump. Coppicing is a traditional practice used for producing wood for fuel, fencing and other purposes. It is particularly effective with species that readily regenerate from stumps, such as willow, hazel and chestnut.

Sustainable cutting practices are often guided by forest management plans, which outline specific goals, methods and timelines for harvesting and regeneration. These plans take into account factors such as the age and structure of the forest, the presence of sensitive habitats or species and the long-term productivity of the land.

Using Wood

Wood is one of the most versatile and valuable resources provided by forests. It is used for a wide range of purposes, from construction and furniture-making to fuel and paper production. Sustainable forest management ensures that wood is harvested in a way that does not compromise the future availability of this resource.

Timber Production: timber is one of the primary products of forest management. It is used for construction, furniture and various industrial applications. Sustainable timber production involves selecting the right species, ensuring proper growth conditions and using harvesting methods that minimize damage to the forest. This also includes practices like replanting or encouraging natural regeneration to replace harvested trees.

Firewood and Fuel: in many parts of the world, wood is a primary source of energy for cooking and heating. Sustainable harvesting of firewood involves cutting dead or dying trees, thinning overstocked stands and using fast-growing species that can quickly replenish. Additionally, using efficient wood stoves or kilns can reduce the amount of wood needed and lower the environmental impact.

Non-Timber Forest Products (NTFPs): forests also provide a wide range of non-timber products, such as fruits, nuts, medicinal plants and resins. These products can be harvested sustainably alongside timber, providing additional income and food resources without depleting the forest. The management of NTFPs requires knowledge of the species involved, their ecological roles and the best practices for sustainable harvesting.

Wood for Craft and Art: beyond commercial and practical uses, wood is also a material for artistic and cultural expression. Crafting items from sustainably sourced wood not only supports traditional skills and cultural heritage but also encourages the conservation of forests by adding value to standing trees.

Conservation and Long-Term Management

The ultimate goal of forest management is to conserve forests for future generations. This involves not only sustainable harvesting practices but also active efforts to protect and restore forest ecosystems. Conservation strategies may include protecting old-growth forests, preserving riparian buffers, restoring degraded areas and managing invasive species.

Long-term forest management requires a commitment to monitoring and adapting practices based on changing conditions. This includes regularly assessing the health of the forest, tracking the success of regeneration efforts and adjusting management plans as needed. It also involves engaging with local communities, governments and other stakeholders to ensure that forest management practices align with broader conservation goals and social values.

Forest management is a complex and multifaceted discipline that requires a deep understanding of ecological principles, sustainable practices and the needs of both human and natural communities. By managing forests responsibly, we can ensure that they continue to provide essential resources, support biodiversity and contribute to the health and well-being of our planet.

Wildlife Management

Wildlife management is an essential aspect of natural resource management that focuses on the sustainable interaction between humans and wildlife. It involves practices that aim to conserve wildlife populations, manage habitats and ensure that wildlife resources are used in a way that benefits both current and future generations. The goal of wildlife management is to balance the needs of wildlife with those of humans, ensuring that ecosystems remain healthy and that species are conserved for their ecological, economic and cultural values.

Monitoring Animal Populations

One of the fundamental components of wildlife management is monitoring animal populations. This process involves collecting data on the number, distribution and health of wildlife species in a particular area. Monitoring is crucial for understanding population dynamics, detecting changes in wildlife communities and identifying potential threats to species.

There are several methods used to monitor wildlife populations, each suited to different types of species and environments:

- **Census and Survey Techniques**: these methods involve counting individuals within a species population to estimate overall numbers. Techniques may include direct counts, where animals are observed and counted in the field or indirect methods, such as counting tracks, scat or nests. Aerial surveys are also used for large, open habitats where ground surveys might be impractical.
- **Radio Telemetry and GPS Tracking**: for more detailed monitoring, wildlife managers often use radio collars or GPS devices attached to animals. These tools allow for the tracking of animal movements, home range size, migration patterns and behavior. This data is invaluable for understanding how animals use their habitat and interact with other species, including humans.
- **Camera Traps**: motion-activated cameras placed in the field can capture images or videos of wildlife, providing data on species presence, abundance and behavior. Camera traps are particularly useful for monitoring elusive or nocturnal species that are difficult to observe directly.
- **Citizen Science**: engaging the public in wildlife monitoring can greatly expand the scope of data collection. Citizen scientists can help by reporting sightings, participating in bird counts or monitoring specific species or habitats. This approach not only enhances data collection but also raises public awareness and involvement in conservation efforts.

Effective population monitoring allows wildlife managers to make informed decisions about conservation priorities, hunting regulations, habitat management and other aspects of wildlife management.

Habitat and Conservation

The conservation of wildlife is inextricably linked to the preservation and management of their habitats. A healthy habitat provides the necessary resources – food, water, shelter and breeding sites – that wildlife need to survive and reproduce. As such, habitat management is a critical focus of wildlife conservation efforts.

Habitat Restoration: in areas where habitats have been degraded by human activity, restoration efforts aim to return these areas to a more natural state. This can involve reforestation, wetland restoration, removal of invasive species and re-establishing natural fire regimes. Restoring habitats not only benefits the species that rely on them but also helps to improve ecosystem services, such as water filtration and carbon sequestration.

Habitat Preservation: protecting existing habitats from further degradation is another key aspect of wildlife management. This might involve the creation of protected areas, such as national parks or wildlife reserves, where human activities are restricted. In these areas, habitats can be preserved in their natural state, providing refuges for wildlife and maintaining biodiversity.

Habitat Connectivity: wildlife often needs to move between different areas to find food, mates or suitable breeding sites. Fragmentation of habitats by roads, agriculture or urban development can impede these movements, leading to isolated populations and reduced genetic diversity. Wildlife corridors, which connect fragmented habitats, are critical for allowing animals to move freely across the landscape. These corridors can be natural features, such as riparian zones or they can be created or maintained through careful planning and management.

Conservation of Critical Habitats: some species require specific types of habitats to survive, such as wetlands for amphibians or old-growth forests for certain bird species. Identifying and protecting these critical habitats is essential for the conservation of those species. This might involve legal protection, land acquisition or the implementation of specific management practices to maintain or enhance habitat quality.

Sustainable Interaction with Wildlife

Managing the interaction between humans and wildlife is another critical component of wildlife management. As human populations grow and expand into natural areas, interactions between humans and wildlife become more frequent, sometimes leading to conflicts. Wildlife management seeks to mitigate these conflicts while ensuring that wildlife populations remain healthy and sustainable.

Human-Wildlife Conflict Mitigation: conflicts between humans and wildlife can arise when animals damage crops, prey on livestock or pose threats to human safety. Wildlife managers work to develop strategies that reduce these conflicts, such as the use of fencing, deterrents or relocation of problematic animals. Education and outreach are also important for helping communities understand and coexist with wildlife.

Sustainable Harvesting: in some cases, wildlife populations are harvested for food, sport or trade. Sustainable harvesting involves setting limits on the number of animals that can be taken, ensuring that populations remain healthy and that the harvest does not threaten the species' long-term viability. This might include regulated hunting seasons, quotas or the use of sustainable hunting practices.

Ecotourism and Recreation: wildlife-based tourism and recreation can provide significant economic benefits to local communities while also supporting conservation efforts. However, it is important that these activities are managed sustainably to avoid negative impacts on wildlife and their habitats. This might involve setting limits on the number of visitors, providing education on responsible wildlife viewing and ensuring that tourism revenue is reinvested in conservation.

Community Involvement: engaging local communities in wildlife management is essential for the success of conservation efforts. Communities often have a deep understanding of local wildlife and ecosystems and their involvement can lead to more effective and sustainable management practices. This might include involving communities in monitoring efforts, developing community-based conservation programs or supporting traditional practices that contribute to wildlife conservation.

Natural resource management is not just about using resources, it's about stewarding them for the future. By understanding the importance of sustainable management, learning how to identify and conserve resources and practicing responsible forest and wildlife management, you can create a system that supports both your needs and the health of the environment. As you implement these practices, remember that every action you take has an impact. Strive to make choices that benefit both you and the natural world, ensuring that these resources remain available for generations to come. Your commitment to natural resource management is a commitment to the future, one where humans and nature coexist in harmony.

Exercise Chapter 15
Mapping and Cataloging Local Natural Resources

Objective: identify and catalog the natural resources available in your local area to better understand your environment and prepare sustainable management strategies.

Materials Needed: local area map (can be paper or digital), pen or pencil, notebook, camera or smartphone for visual documentation, optional adhesive labels.

1. Explore Your Local Area: begin by exploring the territory around you. Identify and take notes on the various natural resources present, such as water sources (rivers, lakes, springs), forests, agricultural areas, soil types, minerals and local wildlife.

2. Catalog the Resources: for each resource identified, note details such as the exact location (you can mark points on the map), approximate quantity, resource quality and current state. For example, if you identify a water source, take note of its flow rate, water quality and signs of pollution.

3. Assess Potential Use: once the resources are cataloged, evaluate how they might be sustainably used. Consider which resources can be exploited without harming the environment, which need conservation or restoration and which might be crucial for long-term self-sufficiency.

4. Document with Photos: use the camera or smartphone to take photos of the identified resources. This will help you visually document the resources and monitor any changes over time.

Deliverable: write a brief report detailing the process of mapping and cataloging local natural resources. Include any challenges you encountered during exploration and how you overcame them. Attach the map with marked points and photos of the resources. Reflect on how the identified resources can be managed sustainably and what actions you might take in the future to preserve them.

Mapping and cataloging local natural resources provides you with a deep understanding of the environment around you and is the first step toward sustainable resource management. This practice not only prepares you to live in harmony with nature but also gives you the tools to protect and enhance the environment for future generations.

Chapter 16
Barter and Trade Techniques

In a world where traditional economic systems may collapse or become unreliable, the ability to barter and trade effectively becomes a critical survival skill. Bartering, the oldest form of commerce, involves the direct exchange of goods and services without using money. This chapter will explore the importance of bartering in crisis situations, the types of items that hold value in such scenarios and the techniques for successful negotiation. Furthermore, it will delve into the creation of trade communities, which can offer a structured environment for bartering and discuss the resources and skills that are most valuable for trade.

In any survival or crisis situation, where currency might lose its value or become scarce, bartering serves as a reliable method of acquiring necessary goods and services. The importance of bartering lies in its simplicity and directness, allowing individuals to exchange what they have for what they need, without the complexities of monetary systems. However, successful bartering requires more than just having items to trade; it demands an understanding of what is valuable in the current context, as well as the ability to negotiate effectively to ensure fair exchanges.

This chapter is structured to provide a comprehensive guide on how to navigate the world of barter and trade, especially in times of crisis. The first section focuses on the basics of bartering, including the importance of this practice in emergencies, identifying valuable barter items and mastering negotiation techniques. The second section covers the steps to create and maintain a trade community, offering insights into community organization, the establishment of rules and guidelines and the overall benefits of bartering within a community framework. The final section discusses the resources and skills that are essential for bartering, emphasizing the importance of having both tangible goods and practical skills to offer in trade, as well as strategies to maximize your bartering potential.

Introduction to Bartering

Bartering, one of the oldest forms of economic exchange, involves trading goods or services directly without the use of money. In today's world, where money and digital transactions dominate the economy, bartering might seem outdated. However, in situations where traditional financial systems break down – such as during natural disasters, economic crises or in remote locations – bartering can become a critical means of acquiring necessary goods and services.

The importance of bartering in crisis situations cannot be overstated. In environments where currency loses its value or becomes inaccessible, the ability to trade tangible goods or skills directly for other needed items can make the difference between survival and hardship. Bartering empowers individuals to leverage what they have to get what they need, fostering a system of mutual aid and resource exchange that can sustain communities through difficult times. This practice becomes especially relevant in scenarios where supply chains are disrupted, inflation skyrockets or banking systems collapse, rendering money ineffective.

In these contexts, certain items become inherently more valuable due to their utility and scarcity. Essentials like food, water, medical supplies, fuel and tools often become the most sought-after goods. Additionally, items that may seem mundane in normal circumstances – such as batteries, candles and hygiene products – can suddenly hold significant bartering power. The value of these items is directly tied to the immediate needs of the community and the availability of alternatives. For example, in a situation where clean water is scarce, water purification tablets or portable filters might be highly prized, far exceeding their regular market value.

Beyond the goods themselves, the art of bartering lies in negotiation. Unlike monetary transactions, where price tags determine value, bartering requires a keen understanding of the relative worth of different items in the current context. Negotiation is a skill that involves assessing the needs of both parties, recognizing the value of what each person brings to the table and finding a mutually beneficial agreement. Effective bartering often hinges on building trust, clear communication and the ability to adapt to changing circumstances. In many cases, the success of a barter deal depends not only on the intrinsic value of the items being exchanged but also on the persuasiveness and interpersonal skills of the traders.

Furthermore, bartering isn't just about the immediate exchange; it often plays a crucial role in establishing ongoing relationships within a community. In times of crisis, when resources are limited, the ability to trade effectively can foster a network of mutual support. Individuals and groups who engage in fair and trustworthy bartering practices are more likely to be seen as reliable partners, which can lead to more favorable trade opportunities in the future.

Bartering also encourages a broader view of what constitutes value. In a barter economy, goods are not the only commodities; skills and services are equally, if not more, important. For example, someone with medical knowledge or the ability to repair tools might trade their expertise for food or other necessities. This highlights the importance of having a diverse set of skills and being adaptable in how you offer them to others. In a functioning barter system, everyone has something to contribute and the focus shifts from the accumulation of wealth to the exchange of value.

Concluding, bartering is a time-tested economic practice that becomes particularly vital in times of crisis. Its importance lies in its flexibility and adaptability, allowing individuals to meet their needs even when traditional financial systems fail. By understanding what items and skills are valuable, mastering negotiation techniques and building strong trade relationships, individuals can navigate the challenges of a barter economy with confidence and resilience. This approach not only ensures survival but also strengthens the bonds within a community, fostering a spirit of cooperation and mutual support that is essential in difficult times.

Creating a Trade Community

Creating a trade community is a vital aspect of sustainable living, particularly in situations where traditional economic systems are disrupted or inaccessible. A well-organized trade community fosters cooperation, mutual support and resilience among its members, enabling them to exchange goods, services and skills effectively. Establishing such a community requires careful planning, clear guidelines and a strong sense of shared purpose. This section delves into the steps necessary to build a thriving trade community, focusing on organization, rules and the overall benefits that bartering brings to the table.

Community Organization

The first step in creating a trade community is organizing the group of individuals or families who will participate. This involves bringing together people with diverse skills, resources and needs, ensuring that the community has a balanced mix of what is available and what is required. The key to successful organization is clear communication and the establishment of a shared vision for the community.

To begin with, it's essential to identify and gather potential members. This can be done through community meetings, social media groups or word of mouth. The goal is to bring together a group that is large enough to provide a variety of goods and services, yet small enough to manage effectively. It's also important to include individuals with leadership qualities who can help facilitate and oversee the functioning of the trade community.

Once the group is established, roles and responsibilities should be defined. These might include roles like a coordinator to organize trade events, a record keeper to track trades and ensure fairness and a mediator to resolve disputes. Defining these roles early on helps the community run smoothly and prevents misunderstandings or conflicts.

Regular meetings are another crucial component of community organization. These meetings provide a forum for members to discuss their needs, offer their resources or services and arrange trades. They also serve as an opportunity to review and adjust the community's rules and practices as needed, ensuring that the system remains fair and efficient.

Rules and Guidelines

Establishing clear rules and guidelines is essential for the smooth operation of a trade community. These rules help to maintain fairness, transparency and trust among members, which are all critical to the success of the community.

One of the first guidelines to establish is the value system for trades. Since bartering does not involve money, the community must agree on how to assess the value of goods and services. This could be done through a point system, where different items and services are assigned a specific number of points based on their perceived value or through direct negotiation where the value is determined case by case. The chosen system should be simple enough for everyone to understand and use consistently.

Another important guideline is setting the frequency and location of trade events. Depending on the size and needs of the community, these could be weekly, bi-weekly or monthly events. Having a regular schedule helps members plan ahead and ensures that everyone has an opportunity to participate. The location should be easily accessible to all members and safe for exchanging goods.

Rules regarding the quality and condition of traded goods are also necessary. To prevent disputes and ensure fairness, the community might establish standards for the condition of items being traded (e.g., goods must be in good working order, food items should be fresh, etc.). It's also beneficial to have a return or exchange policy in place for trades that do not meet agreed-upon standards.

To keep the community strong, it's essential to have a system for resolving disputes. Even in a close-knit community, disagreements can arise over the value of goods or the fairness of a trade. A designated mediator or a small group of trusted individuals can handle these issues, helping to resolve conflicts quickly and fairly without disrupting the community's overall harmony.

Benefits of Bartering

The benefits of establishing a trade community extend far beyond the simple exchange of goods and services. Bartering fosters a sense of community and mutual aid, which can be particularly valuable in times of crisis or economic hardship. When members of a community rely on each other to meet their needs, strong social bonds are formed, creating a support network that enhances overall resilience.

Bartering also promotes sustainability by encouraging the reuse and repurposing of items that might otherwise be discarded. Instead of throwing away surplus items or tools, community members can trade them for something they need, reducing waste and extending the life of resources. This practice not only benefits individual members but also contributes to the community's environmental sustainability.

Moreover, a trade community provides a platform for skill-sharing and education. Members can exchange knowledge and expertise, such as teaching each other how to grow certain crops, repair tools or perform basic medical procedures. This transfer of skills not only empowers individuals but also enriches the entire community by increasing its collective knowledge base.

Another significant benefit of bartering is economic security. In a trade-based community, members are less dependent on traditional monetary systems, which can be volatile or unreliable during economic downturns. By trading goods and services directly, community members can continue to meet their needs even when cash is scarce.

Creating a trade community requires careful planning, clear rules and a commitment to mutual support. By organizing effectively, establishing fair guidelines and recognizing the many benefits of bartering, communities can build a resilient system that not only meets their immediate needs but also strengthens social bonds and promotes long-term sustainability. Whether in times of crisis or as a means to live more sustainably, a well-organized trade community is a powerful tool for fostering cooperation and resilience.

Resources and Skills for Bartering

In a barter economy, where goods and services are exchanged directly without the use of money, the value of resources and skills takes on a new significance. Understanding what resources are valuable, how to acquire and maintain them and which skills are most in demand can greatly enhance your ability to trade effectively. This section delves into the types of resources and skills that hold the highest value in a barter system and offers strategies for leveraging them to your advantage.

Valuable Resources

When it comes to bartering, not all resources are created equal. The value of a particular item depends on its utility, scarcity and demand within the community. Here are some of the most valuable resources that are likely to hold significant bartering power:

- **Food and Water**: in any scenario, food and water are critical for survival, making them among the most valuable resources in a barter economy. Non-perishable food items, such as canned goods, rice, beans and dehydrated foods, are especially sought after. Water, whether in large quantities or purification methods like filters and tablets, is indispensable and highly tradable.
- **Medical Supplies**: first aid kits, bandages, antiseptics, pain relievers and other basic medical supplies are always in high demand, particularly in crisis situations where access to healthcare may be limited. The ability to trade for these items can be life-saving.
- **Fuel and Energy Sources**: in many situations, having access to fuel – whether it's gasoline, propane, wood or even solar power – is crucial for cooking, heating and transportation. These resources are highly valuable in bartering, especially when traditional energy sources are disrupted.

- **Tools and Equipment**: durable tools, from basic hand tools like hammers and wrenches to more specialized equipment like generators or water pumps, hold significant value. These items enable individuals to perform essential tasks and repairs, making them a key asset in trades.
- **Ammunition and Firearms**: in some contexts, particularly in areas where self-defense is a priority, ammunition and firearms can be among the most sought-after resources. However, trading these items requires careful consideration of the legal and ethical implications.
- **Clothing and Shelter Materials**: warm clothing, sturdy boots, blankets and materials for building or repairing shelters are always in demand. These items not only provide immediate comfort but are also critical for long-term survival in harsh environments.

Required Skills

In addition to physical resources, certain skills become incredibly valuable in a barter economy. The ability to offer services or expertise can often be just as valuable – if not more so – than tangible goods. Here are some of the most sought-after skills in a bartering system:

- **Medical and First Aid Skills**: knowledge of first aid, basic medical procedures or even more advanced medical skills can make you an indispensable member of a community. Those with training in emergency medicine, herbal remedies or holistic health practices often find themselves in high demand.
- **Agriculture and Food Production**: skills in gardening, farming, animal husbandry and food preservation are crucial in any self-sufficient community. Individuals who can grow food, care for livestock or preserve harvests through canning, drying or fermenting are invaluable.
- **Mechanical and Repair Skills**: the ability to repair tools, machinery, vehicles or household items is always in demand. Whether you're fixing a broken generator, repairing plumbing or maintaining bicycles, mechanical skills ensure that essential items remain functional.
- **Construction and Carpentry**: those with the ability to build or repair structures – whether it's a simple shelter, a more complex home or even just furniture – are essential in any community. Carpentry, masonry and general construction skills allow for the creation and maintenance of the infrastructure needed for daily living.
- **Hunting, Fishing and Foraging**: in many survival scenarios, the ability to procure food from the environment is critical. Skills in hunting, trapping, fishing and foraging for edible plants can provide a steady food source and make you a valuable trade partner.
- **Communication and Mediation**: while often overlooked, strong communication skills and the ability to mediate disputes are vital in maintaining a harmonious trade community. Those who can negotiate effectively, resolve conflicts and foster cooperation between members play a crucial role in the stability of the barter system.

Trade Strategies

To succeed in a barter system, it's important to develop effective trade strategies that allow you to maximize the value of your resources and skills. Here are some strategies to consider:

- **Diversification**: just as in traditional economies, diversification is key in bartering. Having a variety of resources and skills to offer makes you more adaptable and increases your chances of finding trade opportunities. Diversifying also protects you against the loss or depreciation of any single resource.
- **Networking**: building strong relationships within your trade community is essential. The more connections you have, the more likely you are to hear about trade opportunities and to be trusted as a reliable trade partner. Networking can also help you gain access to rare or high-demand items.
- **Barter Value Assessment**: develop a keen sense of the value of various goods and services within your community. Understanding what others need and what they value most allows you to negotiate better deals. This involves staying informed about the supply and demand dynamics of your community.
- **Negotiation Skills**: being a skilled negotiator is crucial in a barter economy. Learn to recognize when you have leverage in a trade and when it's better to compromise. The goal is to reach a deal that benefits both parties while maintaining fairness and building trust.
- **Stockpiling**: when possible, stockpile high-demand items or resources that have a long shelf life. This strategy allows you to trade when market conditions are favorable, giving you an advantage in negotiations. However, be mindful of the storage conditions to ensure your stockpiles remain usable.

Success in bartering hinges on having valuable resources, developing essential skills and employing strategic trade practices. By understanding what is most needed in your community and honing your ability to provide those goods or services, you can navigate the barter system effectively, ensuring your own needs are met while contributing to the resilience and stability of your community.

Bartering and trade are more than just survival tactics; they are foundational aspects of building resilient communities in times of crisis. By understanding the principles of bartering, learning how to create and maintain trade communities and developing the right resources and skills, you can ensure that you are prepared to thrive even when traditional economic systems fail.

Remember, in a barter economy, your greatest assets are not just the goods you possess, but the skills and relationships you cultivate. Embrace these techniques as a way to strengthen your self-sufficiency and support those around you, creating a network of mutual aid that can weather any storm.

Exercise Chapter 16
Organizing a Community Barter Event

———— ✧✦✧ ————

Objective: organize a community barter event that facilitates the exchange of goods, services and skills among participants, fostering a sense of cooperation and mutual support within your community.

Materials Needed: a location (such as a community center, park or large backyard), tables or blankets for displaying items, signs for labeling sections (e.g., food, tools, skills), a notice board or sign-up sheet, markers, name tags and a basic first aid kit for safety.

Planning the Event: begin by selecting a date and location for the barter event. Ensure the space is large enough to accommodate the number of participants you expect. Promote the event within your community through flyers, social media or word of mouth, encouraging people to bring items or skills they are willing to trade.

Organizing the Layout: arrange the space to clearly mark different categories like food, tools and skills. Set up tables or designated spots for items and provide name tags to facilitate introductions and negotiations.

Establishing Rules: create simple rules to ensure the event runs smoothly. These might include guidelines on fair trading, respecting others' property and maintaining a positive and cooperative atmosphere. Consider setting up a "mediator" role to help resolve any disputes that may arise during trades.

Facilitating Trades: encourage participants to walk around, inspect items and engage in discussions with others. Highlight the importance of clear communication and fair negotiation. If someone is offering a skill or service, suggest they provide a brief demonstration or explanation to potential traders.

Recording Trades: optionally, provide a sign-up sheet or notice board where participants can note successful trades or request future exchanges. This helps build a network of trust and can lead to ongoing bartering relationships within the community.

Deliverable: write a brief report on the outcomes of your barter event, detailing items exchanged, challenges faced and how they were resolved. Include photos and reflect on how the event strengthened community ties and suggest improvements.

Organizing a community barter event not only helps meet immediate needs but also fosters a sense of solidarity and resilience within your community. It encourages participants to recognize the value of their own resources and skills while building lasting relationships that can support everyone in times of need.

Chapter 17
Disaster Preparedness

In an unpredictable world, disaster preparedness is not just a precaution, it is a necessity. Whether facing the threat of earthquakes, floods or other natural disasters, being prepared can mean the difference between life and death. Effective disaster preparedness involves understanding the types of disasters you may face, creating comprehensive emergency plans and ensuring that every member of your family knows their role in these plans. This chapter will guide you through the essential steps of disaster preparedness, focusing on the specific measures you can take to protect yourself and your loved ones from earthquakes and floods.

Disasters can strike at any time, often without warning. The best way to protect yourself and your family is to be proactive in your preparations. This includes not only gathering the necessary supplies but also developing a mindset that prioritizes safety and quick action. Preparedness is about more than just stocking up on food and water, it involves understanding the risks you face, creating detailed plans and practicing those plans until they become second nature.

This chapter is divided into three sections. The first section provides an overview of disaster preparedness, including the types of natural disasters you may encounter, the importance of emergency planning and how to create a family plan that everyone can follow. The second section focuses on earthquake preparedness, offering detailed advice on safety measures, creating emergency kits and survival techniques specific to seismic events. The final section covers flood preparedness, discussing prevention measures, building barriers and the critical steps to take during evacuation and to ensure safety during a flood.

Introduction to Disaster Preparedness

Disaster preparedness is a critical component of ensuring safety and survival in the face of natural and human-made disasters. It involves a comprehensive approach to identifying potential threats, planning responses and equipping individuals and communities with the tools and knowledge needed to mitigate the impact of these events. Whether it's an earthquake, flood, hurricane or wildfire, being prepared can significantly reduce the risks and increase the chances of survival.

The need for disaster preparedness has become increasingly evident as the frequency and intensity of natural disasters continue to rise globally. Climate change, urbanization and population growth have all contributed to making communities more vulnerable to disasters. Therefore, understanding the types of disasters that could affect you and knowing how to prepare for them, is more important than ever.

Types of Natural Disasters: natural disasters come in many forms, each with its own set of challenges. Earthquakes, for example, strike suddenly and without warning, causing widespread damage and posing severe risks to human life. Floods, on the other hand, might be anticipated with some notice, but they can rapidly devastate large areas, disrupting lives and displacing communities. Hurricanes combine high winds and heavy rains, leading to both immediate destruction and long-term challenges like flooding and infrastructure damage. Other disasters include tornadoes, wildfires, landslides and volcanic eruptions, each requiring specific preparedness strategies tailored to their unique characteristics. Recognizing the types of natural disasters that are most likely to occur in your region is the first step in effective disaster preparedness.

Emergency Planning: at the heart of disaster preparedness is emergency planning. A well-thought-out emergency plan outlines the steps to take before, during and after a disaster strikes. This plan should include how to secure your home, how to communicate with family members and where to go if evacuation becomes necessary. Emergency plans should be detailed and include contingencies for various scenarios, such as what to do if communication networks fail or if evacuation routes are blocked. It's also important to regularly review and update your emergency plan, ensuring that all family members are familiar with it and know their roles.

A key element of emergency planning is assembling an emergency kit that contains essential supplies like food, water, medications and tools that can sustain you and your family for at least 72 hours. This kit should be easily accessible and portable in case you need to evacuate quickly. Additionally, your plan should include a communication strategy, such as using a designated family member as a point of contact or agreeing on a meeting place if you're separated during the disaster.

Creating a Family Plan: a family disaster plan is a personalized strategy that outlines how each member of the family will respond in an emergency. This plan should account for everyone's needs, including children, elderly family members and pets. It should also include special considerations, such as how to handle medical conditions or disabilities during a disaster.

Start by identifying potential hazards in your home and neighborhood. For example, if you live in an earthquake-prone area, ensure that heavy furniture is secured to walls and that everyone knows to "Drop, Cover and Hold On" during a quake. If flooding is a concern, make sure everyone knows the best routes to higher ground.

Communication is a critical part of your family plan. Establish a method for contacting each other during a disaster, whether through phone calls, texts or a pre-arranged meeting place. It's also wise to have an out-of-town contact that all family members can reach if local communications are disrupted. Practicing your family plan through drills will help ensure that everyone knows what to do when a real disaster occurs.

Disaster preparedness is about more than just reacting to emergencies, it's about anticipating them and taking proactive steps to protect yourself and your loved ones. By understanding the types of disasters that could affect you, creating detailed emergency plans and preparing your family, you can face these challenges with confidence. Remember, preparedness is an ongoing process that requires regular review and adjustment. The time and effort you invest in preparing today can make all the difference in a crisis, helping to safeguard your life and the lives of those you care about.

Earthquake Preparedness

Earthquake preparedness is an essential aspect of disaster planning, especially for those living in regions prone to seismic activity. Earthquakes can strike without warning, causing widespread damage and potentially life-threatening situations. Being well-prepared can significantly reduce the risk of injury and improve your chances of survival during and after an earthquake. This section will cover the key elements of earthquake preparedness, including safety measures, the creation of emergency kits and essential survival techniques.

Safety Measures

Safety measures are the first line of defense in earthquake preparedness. These precautions are designed to protect you and your property before, during and immediately after an earthquake. Implementing these measures can minimize the risk of injury and damage:

- **Home Preparation**: begin by securing your living space. Heavy furniture, such as bookshelves, cabinets and appliances, should be anchored to walls to prevent them from toppling over during the shaking. Install latches on cabinet doors to keep them from flying open and causing items to spill out. Ensure that pictures, mirrors and other wall hangings are securely fastened to prevent them from falling. In the kitchen, store heavy items on lower shelves to reduce the risk of them falling and causing injury.
- **Structural Integrity**: if you live in an earthquake-prone area, it's crucial to assess the structural integrity of your home. Have a professional check that your house is built to withstand seismic activity and consider retrofitting it if necessary. Retrofitting might include reinforcing the foundation, securing the house to the foundation or adding braces to support the structure.

- **Developing Safe Zones**: identify safe zones within your home where you can take cover during an earthquake. These are typically areas away from windows, glass and heavy objects that could fall. The safest places are usually under sturdy furniture like tables or against an interior wall away from potential hazards. Make sure every member of your household knows where these safe zones are and practices moving to them quickly during earthquake drills.
- **Regular Drills**: conducting regular earthquake drills is essential for ensuring that everyone in your household knows what to do when an earthquake strikes. Practice the "Drop, Cover and Hold On" technique, which involves dropping to your hands and knees, covering your head and neck under a sturdy piece of furniture or near an interior wall and holding on until the shaking stops. These drills can be life-saving, helping to instill quick, automatic responses during an actual event.

Creating Emergency Kits

An emergency kit is a critical component of earthquake preparedness. This kit should contain all the essentials needed to survive for at least 72 hours following an earthquake. Since earthquakes can disrupt services and isolate communities, having a well-stocked emergency kit can make a significant difference in your ability to cope with the aftermath.

Your emergency kit should include the following basic supplies:
- **Water**: at least one gallon per person per day for drinking and sanitation, stored in sealed, durable containers.
- **Food**: non-perishable food items that are easy to prepare, such as canned goods, dried fruits, nuts and energy bars. Include a manual can opener if needed.
- **First Aid Supplies**: a comprehensive first aid kit with bandages, antiseptics, pain relievers, prescription medications and other necessary medical supplies.
- **Tools and Equipment**: flashlights with extra batteries, a multi-tool, a whistle to signal for help, dust masks to filter contaminated air and a wrench or pliers to turn off utilities.
- **Personal Items**: include warm clothing, sturdy shoes, blankets and personal hygiene items such as soap, hand sanitizer and feminine hygiene products.
- **Communication Devices**: a battery-powered or hand-crank radio to stay informed about emergency information, as well as fully charged cell phones with backup power banks.
- **Important Documents**: copies of identification, insurance policies, medical records and any other critical documents should be stored in a waterproof container.
- **Personalization**: tailor your emergency kit to your specific needs. For example, if you have pets, include pet food, water and any necessary medications. If you have infants or elderly family members, ensure that you have baby formula, diapers and other essential items. Consider the unique needs of each member of your household and adjust your kit accordingly.
- **Accessibility**: store your emergency kit in a location that is easily accessible and known to all family members. Since earthquakes can make doors difficult to open or obstruct pathways, keep the kit in a place where you can reach it without having to move heavy objects or navigate dangerous areas.

Consider having smaller kits in multiple locations, such as in your car or workplace, to ensure you are prepared wherever you are.

Survival Techniques

Knowing how to survive during and immediately after an earthquake is crucial. While no one can predict when an earthquake will happen, being prepared with the right techniques can save lives.

When the ground starts shaking, it's vital to act quickly:
- **Indoors**: follow the "Drop, Cover and Hold On" method. Drop to your hands and knees to prevent being knocked over, cover your head and neck with your arms or seek shelter under a sturdy piece of furniture and hold on until the shaking stops. Stay away from windows, mirrors and exterior walls. Do not attempt to run outside as this can increase your risk of injury from falling debris.
- **Outdoors**: move to an open area away from buildings, streetlights, trees and anything else that could fall. Drop to the ground and protect your head and neck. Stay in place until the shaking stops, avoiding the temptation to run indoors or into more confined areas.
- **In a Vehicle**: if you are driving when an earthquake strikes, pull over to a safe location away from overpasses, bridges and power lines. Remain in the vehicle with your seatbelt fastened until the shaking stops. Once it's safe, proceed cautiously, being aware of potential road damage or debris.
- **After the Earthquake**: the period following an earthquake can be just as dangerous as the event itself. Be prepared for aftershocks, which can occur minutes, hours or even days after the initial quake. Aftershocks can cause further damage to weakened structures, so it's important to stay vigilant.
- **Checking for Injuries**: first, check yourself and others for injuries. Administer first aid if necessary and avoid moving seriously injured individuals unless they are in immediate danger.
- **Inspecting for Hazards**: inspect your surroundings for hazards such as gas leaks, damaged electrical wires and structural damage. If you smell gas, turn off the gas supply, open windows and evacuate the area. Do not use matches, lighters or electrical switches, as they could ignite a fire.
- **Communicating**: use your emergency communication devices to stay informed and to contact family members or emergency services. Be mindful of conserving battery power by using devices only when necessary.
- **Evacuation**: if your home is no longer safe to stay in due to structural damage or other hazards, evacuate to a pre-planned safe location. Follow local authorities' instructions regarding evacuation routes and shelter locations.

Earthquake preparedness is not just about surviving the immediate shaking but also about ensuring long-term safety and resilience. By implementing these safety measures, creating a well-equipped emergency kit and mastering survival techniques, you can significantly increase your chances of protecting yourself and your loved ones during and after an earthquake. Remember, preparation is an ongoing process that requires regular updates and practice.

Flood Preparedness

Flood preparedness is crucial for minimizing damage and ensuring safety when faced with the threat of rising waters. Floods can occur suddenly due to heavy rainfall, storm surges or rapid snowmelt, often with little warning. Being well-prepared can make a significant difference in your ability to protect your property and, more importantly, ensure your family's safety. This section covers prevention measures, techniques for building barriers and guidelines for evacuation and safety during a flood.

Prevention Measures

Flood prevention measures are your first line of defense against water damage and can greatly reduce the impact of a flood on your property and community. These measures focus on protecting your home, managing water flow and staying informed about potential flooding risks.

Home Protection: begin by ensuring that your home is as flood-resistant as possible. This involves several structural improvements and maintenance tasks:

Elevation: if you live in a flood-prone area, one of the most effective ways to protect your home is to elevate it above the expected flood level. This might involve raising the entire structure or, more commonly, elevating critical utilities like electrical panels, water heaters and HVAC systems above the base flood elevation.

Flood-Proofing: seal the foundation and basement walls with waterproof coatings or barriers to prevent water from seeping in. Installing sump pumps with battery backups can help remove water that does enter your basement. Consider installing flood vents that allow water to flow through your foundation rather than build up pressure that could cause structural damage.

Drainage Systems: ensure that your property has an effective drainage system. Clean gutters and downspouts regularly to prevent water from pooling around your home. Direct downspouts away from your foundation and toward areas where water can safely drain away. Additionally, consider installing French drains or other landscaping features that help divert water away from your home.

Community-Level Prevention: beyond individual actions, flood prevention also involves community-level efforts. Advocate for proper land use planning that discourages development in flood-prone areas and supports the preservation of wetlands and natural flood plains, which act as natural buffers against floods.

Stormwater Management: work with local authorities to ensure that the community has adequate stormwater management systems, such as retention basins, levees and floodwalls. These structures are designed to control water flow and prevent it from overwhelming urban areas during heavy rains.
Staying Informed: One of the most critical aspects of flood preparedness is staying informed about potential flood risks. Sign up for local emergency alerts and weather updates. Know the flood risk level of your area,

which can usually be found through local government resources or the Federal Emergency Management Agency (FEMA) in the United States.

Building Barriers

When a flood is imminent, building barriers can be an effective way to protect your property from encroaching water. These barriers range from simple sandbag walls to more advanced and permanent flood defenses.

Sandbags: sandbags are one of the most common and accessible methods for creating a flood barrier. Properly filled and placed sandbags can redirect water flow away from your home and prevent floodwaters from entering doors and low windows.

Filling and Placement: fill sandbags about halfway full and fold the top over; this allows them to be more flexible and stack tightly together. Place them tightly in a staggered pattern, similar to laying bricks, to create a stable barrier. The wall should be at least two feet high to be effective, but this may need to be higher depending on expected flood levels.

Water-Filled Barriers: an alternative to traditional sandbags is the use of water-filled barriers. These are large, flexible tubes that are filled with water to create a heavy, impermeable barrier against floodwaters. They are quicker to deploy than sandbags and can be reused, making them an efficient choice for larger properties or repeated use.

Temporary Flood Walls: for more substantial protection, consider installing temporary flood barriers or flood gates around doors, windows and other entry points. These barriers can be quickly assembled before a flood and disassembled afterward. Some systems use interlocking panels or hinged gates that can be swung into place when needed.

Permanent Flood Defenses: in areas where flooding is a regular threat, it may be worth investing in more permanent solutions like levees, berms or floodwalls. These structures are designed to provide long-term protection and can be integrated into the landscape in a way that enhances the property's overall resilience to floods.

Internal Barriers: if water does breach your external defenses, internal barriers like inflatable door dams or water-activated flood bags can provide a last line of defense within your home. These are placed at the entrances to rooms or on stairways to prevent water from spreading throughout the interior.

Evacuation and Safety

Evacuation is often the safest option when a flood is imminent or ongoing. Having a clear evacuation plan in place can save lives and ensure that you and your family can reach safety quickly and efficiently.

Your evacuation plan should include several key elements:
- **Evacuation Routes**: identify multiple routes out of your area, as some roads may become impassable during a flood. Familiarize yourself with these routes and practice them if possible. Include both walking and driving routes in case vehicles are not usable.
- **Meeting Points**: establish a safe meeting point where all family members can reunite if separated. This could be a designated spot within your neighborhood or a specific location outside of the flood-prone area.
- **Emergency Contacts**: compile a list of emergency contacts, including local emergency services, family members and friends who live outside the flood zone. Ensure everyone in your household has access to this list, either digitally or on paper.
- **During the Flood**: if you are ordered to evacuate, do so immediately. Delaying evacuation can put you in greater danger as roads become flooded or blocked. Listen to local authorities for instructions on which routes to take and where to find shelter.
- **Turn Off Utilities**: before leaving, turn off all utilities, including electricity, gas and water. This can help prevent further damage to your home and reduce the risk of fire or gas leaks.
- **Avoid Floodwaters**: never attempt to walk, swim or drive through floodwaters. Just six inches of moving water can knock over an adult and one foot of water can sweep away a vehicle. Floodwaters may also be contaminated with sewage, chemicals or debris, posing serious health risks.
- **After the Flood**: once the floodwaters recede, it's important to return home only when authorities have declared it safe to do so. Be cautious of potential hazards such as structural damage, contaminated water or mold growth.
- **Documenting Damage**: upon returning, document any damage to your property with photos and notes. This will be important for insurance claims and any disaster assistance you may need to apply for.
- **Cleaning Up Safely**: wear protective clothing, including gloves, boots and masks, when cleaning up after a flood. Disinfect surfaces that have come into contact with floodwater and be mindful of the potential for mold growth.

Flood preparedness requires thorough planning, from preventive measures and building barriers to having a well-practiced evacuation plan. By taking these steps, you can protect your home, ensure your family's safety and minimize the impact of floods on your life. Remember, the key to effective flood preparedness is to act well before the waters start to rise.

Disaster preparedness is about being ready for the unexpected. Whether it's an earthquake, flood or any other type of disaster, the steps you take now can protect your family and increase your chances of survival. By understanding the risks, creating detailed plans and practicing those plans, you can face any disaster with confidence and resilience. The safety and well-being of your family depend on your readiness, so take the time to prepare now. Your efforts will not only safeguard your loved ones but also contribute to the resilience and recovery of your entire community.

Exercise Chapter 17
Creating a 72-Hour Emergency Kit

Objective: assemble a 72-hour emergency kit tailored to your personal needs, ensuring that you are prepared for any disaster situation.

Materials Needed: large, durable backpack or storage container, non-perishable food items, water, first aid kit, flashlight, extra batteries, multi-tool, hygiene products, clothing, blanket, important documents, communication devices and family members for input.

1. Research Essential Items: begin by researching the essential items that should be included in a 72-hour emergency kit. Focus on gathering enough food, water and medical supplies to last at least three days. Include items such as canned goods, energy bars, bottled water, a first aid kit and personal medications.

2. Customize Your Kit: consider the specific needs of your household. If you have children, include items such as baby formula, diapers and comfort toys. For pets, pack food, water and any necessary medications. Make sure to include hygiene products like soap, hand sanitizer and feminine hygiene products, as well as extra clothing and blankets for warmth.

3. Prepare Important Documents: gather important documents such as identification, insurance policies, medical records and emergency contact information. Store these in a waterproof container within your kit. Ensure that copies of these documents are accessible to all family members.

4. Assemble and Store Your Kit: pack all items into a large, durable backpack or storage container that is easy to carry in case of evacuation. Store the kit in an easily accessible location known to all family members. Consider placing smaller versions of the kit in your car or workplace.

Deliverable: write a brief report on the process of assembling your 72-hour emergency kit. Include a list of the items you packed, any challenges you faced and how you overcame them. Attach photos of the completed kit. Reflect on how well-prepared you feel after creating the kit and any improvements you might consider for the future.

Creating a 72-hour emergency kit is a crucial step in disaster preparedness, providing peace of mind and practical support in the event of an emergency. Regularly updating and reviewing your kit ensures that you remain prepared and ready to face any unexpected situations.

Chapter 18
Navigation and Orientation Techniques

In an age dominated by GPS and digital maps, the importance of traditional navigation skills might seem diminished. However, these skills remain crucial, especially in survival situations or off-grid adventures where modern technology may fail. The ability to navigate without relying on electronic devices can be a life-saving skill, enabling you to find safety and maintain direction even in the most challenging environments. Understanding how to use natural landmarks, the sun and the stars, alongside mastering tools like maps and compasses, is essential for anyone venturing into the wilderness or unfamiliar territories.

This chapter explores the timeless art of navigation, highlighting the importance of orientation and the tools and techniques that ensure you always know where you are and how to reach your destination.

Introduction to Navigation

Navigation is the practice of determining your position and plotting a course to reach a specific location. Throughout history, this skill has been essential for explorers, travelers and anyone needing to traverse unknown landscapes. In our modern world, while digital tools have become commonplace, traditional navigation methods retain their value, particularly when electronic devices are unavailable or unreliable. Mastering the basics of orientation – knowing your position relative to your surroundings – is the foundation of effective navigation. This skill involves using tools like maps and compasses, as well as natural cues such as the sun and stars, to maintain direction and reach your destination safely. Whether you're in a dense forest, a vast desert or navigating urban streets during an emergency, the ability to orient yourself and navigate efficiently remains a vital skill.

Orientation is the fundamental aspect of navigation. It involves knowing your location relative to your surroundings, including the directions (north, south, east and west) and key landmarks or features that can guide your path. Without a clear sense of orientation, even the most detailed map or the most advanced compass can become useless. Proper orientation ensures that you are moving in the right direction and allows you to make informed decisions about your route, especially in unfamiliar environments. Whether navigating through a dense forest, across a desert or within an urban landscape, maintaining orientation is essential to reaching your destination safely and efficiently.

The Importance of Navigation in Various Contexts

Navigation is not only vital for wilderness survival but also plays a critical role in everyday situations. For hikers and adventurers, good navigation skills prevent them from getting lost in remote areas where rescue could be difficult. For sailors, pilots and mariners, precise navigation is crucial to avoid hazards and ensure safe passage across vast bodies of water. Even in urban environments, where GPS systems are commonly used, understanding how to navigate without technology can be invaluable during emergencies, such as natural disasters, where electronic systems might fail.

In military and survival scenarios, navigation is a life-saving skill. Soldiers and survivalists often train extensively in land navigation, learning to move through challenging terrains using maps, compasses and natural signs. This training ensures they can operate effectively even when GPS is unavailable or unreliable. For explorers and travelers, mastering navigation allows for greater independence and confidence when venturing into unknown regions.

The Evolution and Tools of Navigation

Navigation has evolved significantly over the centuries. Early navigators relied on the stars, the sun and landmarks to guide their way. The invention of the magnetic compass in the 12th century revolutionized navigation, providing a reliable way to determine direction even in poor visibility or featureless environments. With the development of cartography, maps became essential tools, offering detailed representations of the land and sea, complete with distances, elevations and geographic features.

Today, while GPS and other digital navigation systems have become the norm, traditional tools like the compass and map remain indispensable, especially in scenarios where technology fails. These tools require no batteries, are unaffected by signal issues and provide a deeper understanding of the landscape. Learning to use them effectively involves more than just technical know-how; it requires an understanding of geography, topography and environmental patterns.

The Fundamentals of Effective Navigation

Effective navigation combines knowledge of tools with an understanding of the environment. Key techniques include reading and interpreting maps, using a compass to determine and maintain direction and planning routes that account for terrain, obstacles and potential hazards. In addition, natural navigation methods – such as using the sun, stars and natural signs like wind patterns, water flow and vegetation – complement these tools and provide backup options when equipment is unavailable.

Being able to navigate effectively also involves situational awareness, the ability to constantly observe and interpret your surroundings. This skill helps in recognizing landmarks, understanding the implications of changes in the environment and adjusting your route as needed. For example, changes in weather, such as fog or rain, can obscure visibility and landmarks, requiring adjustments in your navigation strategy.

In summary, navigation is a critical skill that extends beyond the use of tools. It involves a deep understanding of the environment, the ability to stay oriented and the knowledge to use both traditional and natural methods to find your way. Whether you are a seasoned explorer, a weekend hiker or someone preparing for emergencies, mastering navigation ensures that you can confidently and safely reach your destination, regardless of the challenges you face.

Using a Compass and Maps

Navigating using a compass and map is an essential skill that forms the backbone of traditional navigation techniques. Whether you're hiking through dense forests, crossing deserts or exploring unfamiliar terrains, the ability to use these tools effectively can be the difference between reaching your destination and getting hopelessly lost. This section will delve into the key elements of map reading, compass use and route planning, providing you with the knowledge and confidence to navigate any landscape.

Map Reading

Understanding how to read a map is the first step in mastering navigation. Maps are rich with information, representing the physical features of the terrain, such as mountains, valleys, rivers and man-made structures like roads and buildings. Learning to interpret these symbols and translate them into real-world understanding is critical for effective navigation.

Types of Maps: different types of maps serve different purposes. Topographic maps are among the most commonly used in outdoor navigation, providing detailed information about the elevation and contour of the land. Contour lines on these maps represent changes in elevation, with closer lines indicating steeper terrain and wider spaces suggesting gentler slopes. Road maps, on the other hand, focus on transportation routes, while orienteering maps are specifically designed for competitive navigation and include intricate details about the terrain.

Map Symbols and Legends: every map includes a legend that explains the symbols and scale used. Common symbols might represent trails, campsites, water sources and natural features. The scale of the map is also crucial, as it indicates the ratio of a distance on the map to the actual distance on the ground, allowing you to estimate distances and plan your route accordingly.

Orienting the Map: before you can use a map effectively, it must be oriented correctly. This means aligning the map with the surrounding terrain so that the features on the map correspond to those in your environment. This is typically done by aligning the map with the north as indicated by your compass. Once oriented, the map becomes a powerful tool for visualizing your surroundings and planning your route.

Using a Compass

A compass is a simple yet indispensable tool for determining direction. Combined with a map, it allows for precise navigation across any landscape. Here's how to use a compass effectively:

- **Basic Components**: a standard compass includes a magnetic needle that always points toward magnetic north, a rotating bezel marked with degrees from 0 to 360 and a baseplate that often includes a ruler for measuring distances on maps. Some compasses also feature a sighting mirror or a clinometer for measuring slopes.
- **Finding North**: to use the compass, start by holding it flat in your hand, ensuring that the needle can rotate freely. The red end of the needle points toward magnetic north. By rotating the bezel so that its "N" marking aligns with the needle, you can determine the direction you're facing.
- **Taking a Bearing**: taking a bearing means determining the direction from your current location to a distant object or point on a map. To do this, point the direction-of-travel arrow on the compass toward your target, then rotate the bezel until the needle aligns with the north marking on the compass. The degree reading at the index line gives you the bearing. This bearing can then be followed to navigate directly to your destination.
- **Using a Compass with a Map**: to combine map and compass navigation, first place the map on a flat surface and orient it to the terrain using the compass. Next, identify your current location on the map and the location you wish to reach. Draw a straight line between these two points. Place the edge of the compass along this line with the direction-of-travel arrow pointing toward your destination. Rotate the bezel until the needle is aligned with the north marking on the map. The degree reading on the compass now represents the bearing you need to follow.

Route Planning

Planning your route is a crucial step in any journey. It involves more than just selecting the shortest path; you must consider the terrain, the difficulty of the journey and the time required to reach your destination.

Assessing the Terrain: start by studying the topography on your map. Look for natural obstacles like rivers, cliffs or dense forests that could make your route more challenging. Consider alternative paths that might be longer but easier to traverse.

Distance and Time Estimation: using the map's scale, estimate the distance between key points on your route. Factor in the terrain's difficulty, as steep ascents or rough ground will slow your progress. It's also essential to plan for breaks and ensure you have enough daylight to complete the journey safely.

Marking Waypoints: identify and mark key waypoints along your route. These are points where you can check your position, adjust your course or take a break. Waypoints might include prominent landmarks, trail intersections or natural features like a large rock or a bend in a river.

Contingency Planning: always plan for the unexpected. Identify alternative routes in case your primary path becomes impassable due to weather, obstacles or other unforeseen events. Make sure you have enough supplies to handle delays and know how to return to your starting point if necessary.

Using a compass and map is an invaluable skill for anyone venturing into the wilderness or exploring unfamiliar territory. By mastering map reading, compass use and route planning, you can navigate with confidence, knowing that you have the tools and knowledge to find your way. These skills not only enhance your safety but also open up a world of exploration, allowing you to venture into remote areas with the assurance that you can always find your way back.

Natural Navigation

Natural navigation is the art of finding your way without relying on conventional tools like maps and compasses. Instead, it involves using environmental cues – such as the sun, stars, plants and terrain – to determine direction and location. This skill has been used by indigenous cultures and explorers for millennia and remains invaluable in situations where modern tools are unavailable or fail. This section explores the key techniques of natural navigation, including orienting with the sun and stars, recognizing natural signs and employing survival techniques to navigate through the wilderness.

Orienting with the Sun and Stars

One of the most reliable natural navigation methods involves using celestial bodies, the sun during the day and stars at night.

Using the Sun: the sun is a consistent guide throughout the day. In the Northern Hemisphere, it rises in the east and sets in the west. At noon, it will be due south and in the Southern Hemisphere, it will be due north. By observing the sun's position at different times of the day, you can approximate your cardinal directions. For a more precise method, the shadow stick technique can be employed. Place a stick upright in the ground and mark the tip of its shadow with a small object or stone. After about 15 minutes, mark the tip of the shadow again. Draw a line between the two marks, this line will run approximately east to west, with the first mark indicating west and the second mark east.

Using the Stars: at night, the stars offer a dependable method for orientation. The North Star or Polaris, is located nearly directly over the North Pole and can be found by locating the Big Dipper constellation. The two stars at the edge of the Big Dipper's bowl, known as the "Pointer Stars," always point to the North Star. In the Southern Hemisphere, finding direction is a bit trickier as there is no equivalent of Polaris. However, the Southern Cross constellation can be used. By extending an imaginary line along the long axis of the cross, it will point roughly toward the South Pole.

Recognizing Natural Signs

Beyond celestial navigation, the natural world is full of subtle indicators that can help you find your way. These signs can be found in the behavior of plants, the flow of water and the general layout of the terrain.

Vegetation and Growth Patterns: plants can provide clues about direction and environmental conditions. For instance, in the Northern Hemisphere, trees often have more foliage and growth on their southern side due to greater sun exposure. Moss, on the other hand, commonly grows on the northern side of trees, rocks and structures because it prefers damp, shaded environments. However, these are general rules and can be influenced by local conditions, so they should be used in conjunction with other navigation methods.

Waterways and Drainage Patterns: water flows downhill and typically follows the path of least resistance, collecting in streams, rivers and lakes. Understanding the local watershed can help you orient yourself, as major rivers often flow toward the ocean or a large body of water. If you find a stream, following it downstream will usually lead to larger rivers and eventually settlements. Additionally, in many regions, settlements are often located near water sources, so locating a river or stream can lead you to civilization.

Animal Behavior: wildlife can also offer navigational hints. Birds often fly toward water sources in the morning and evening and migratory birds follow specific routes that can indicate direction. Ants tend to build their nests on the warmer, sunnier side of trees or objects, which in the Northern Hemisphere is typically the south side. Observing these behaviors can provide additional clues about your orientation.

Survival Techniques

When navigating without tools, survival techniques often overlap with natural navigation methods. These strategies not only help you find your way but also ensure your safety and well-being in the wilderness.

Maintaining Direction: when navigating by natural signs, it's crucial to maintain a consistent direction, especially in dense forests or deserts where visibility is limited. A simple method is to choose a distant landmark in your desired direction and walk toward it. Upon reaching it, select another landmark and continue. This technique prevents you from unintentionally walking in circles, which is a common risk when visibility is poor or when traveling through homogenous terrain.

Pacing and Timing: estimating distances and timing your travel are essential survival skills. By counting your steps (pacing) or noting the time taken to walk a known distance, you can keep track of how far you've traveled. This can be particularly useful when retracing your steps or estimating how long it will take to reach a particular location, especially when following natural signs or celestial cues.

Using the Moon for Navigation: the moon can also serve as a navigational aid, especially on clear nights. When the moon is visible, you can approximate cardinal directions by observing its position relative to the time of night. In the Northern Hemisphere, the moon will generally appear in the eastern sky after sunset, move across the southern sky and set in the west by dawn. Understanding this pattern can help you maintain your course when stars are obscured or other landmarks are not visible.

Adapting to Changing Conditions: natural navigation requires adaptability, as environmental conditions can change rapidly. Weather, for instance, can obscure celestial bodies or alter the landscape. In such cases, it's important to rely on multiple methods of navigation rather than depending on a single cue. For example, if clouds obscure the sun, moon or stars, you might shift focus to natural signs like vegetation patterns or the direction of water flow.

Natural navigation is both an art and a science, rooted in keen observation and a deep understanding of the environment. By mastering techniques such as orienting with the sun and stars, recognizing natural signs and applying essential survival strategies, you can navigate confidently even without modern tools. These skills not only enhance your ability to explore and survive in the wilderness but also connect you with ancient practices that have guided humanity for thousands of years. In situations where technology fails, these time-tested methods ensure that you can always find your way, using the natural world as your guide.

Navigation and orientation are fundamental survival skills that empower you to explore new places, overcome challenges, nand find your way in the most difficult situations. Whether you are using a compass and map or relying on natural navigation techniques, the ability to determine your direction and stay on course is essential for your safety and success. By mastering these skills, you not only increase your confidence in the wilderness but also ensure that you can lead yourself and others to safety when it matters most.

Remember, the key to effective navigation is preparation, practice and a keen awareness of your surroundings. Keep honing these skills and you'll always know where you are and where you're going, no matter where life's journey takes you.

Exercise Chapter 18
Navigating Using Only a Compass and Map

Objective: successfully navigate a designated outdoor area using only a compass and a map, demonstrating your ability to orient yourself, take bearings and follow a planned route.

Materials Needed: compass, topographic map of the area, notebook, pencil, ruler and a watch.

1. Study the Map and Plan Your Route: thoroughly examine the map of your area. Identify key landmarks, contour lines and potential obstacles. Use the map's scale to estimate distances and plan your route from your starting point to a predetermined destination. Mark your route and note important bearings and waypoints.

2. Orient the Map and Take Bearings: in the field, use your compass to orient the map to the terrain. Determine the first bearing by aligning the compass with your route on the map and rotating the housing until the needle points north. The direction-of-travel arrow will indicate the bearing.

3. Follow the Route Using the Compass: navigate by following the bearing on your compass. Walk toward the first waypoint, checking your surroundings against the map to stay on course. Adjust as needed, using both your compass and map.

4. Navigate to the Final Destination: continue following your planned route, taking new bearings at each waypoint and adjusting for obstacles or changes in terrain. Use your map to verify your position regularly.

5. Return and Reflect: after reaching your destination, use the same techniques to navigate back. Review your navigation process, noting challenges and how you resolved them.

Deliverable: write a brief report detailing your navigation experience using the compass and map. Include the route you planned, any deviations you made and how you resolved them. Attach photos of your map, compass and key points along your route. Reflect on what you learned and how you could improve your navigation skills in the future.

Navigating with just a compass and map hones your understanding of traditional navigation techniques, building confidence and skills for exploring unfamiliar terrain without relying on modern technology.

Chapter 19
Managing Health Emergencies

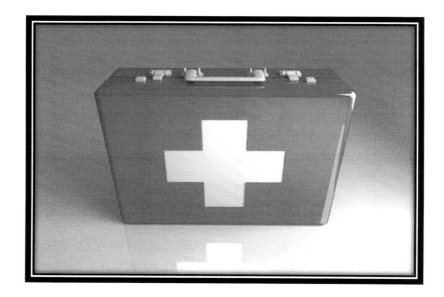

In any survival situation, the ability to manage health emergencies is not just an advantage, it's a necessity. Whether you're dealing with minor injuries, severe wounds or life-threatening diseases, the ability to respond quickly and effectively can make the difference between life and death. This chapter will equip you with the knowledge and skills needed to handle various health emergencies, focusing on medical preparedness, wound treatment and managing diseases and infections with both conventional and natural remedies.

When you find yourself in a survival scenario, access to professional medical help might be limited or non-existent. This reality underscores the importance of being well-prepared to handle health emergencies on your own. Medical preparedness goes beyond having a well-stocked first aid kit; it involves understanding the types of emergencies you might encounter, knowing how to treat them and having a solid plan that everyone in your group understands and can execute. Effective management of health emergencies requires not just the right tools, but also the knowledge and foresight to use them correctly and efficiently.

This chapter is divided into three key sections. The first section covers the importance of medical preparedness, identifying common emergencies you may face and the process of creating a comprehensive emergency plan. The second section delves into the treatment of wounds, providing detailed insights into different types of wounds, suturing and bandaging techniques and methods for preventing infections. The final section addresses the identification and treatment of common diseases and infections, including the use of natural remedies when conventional medicine is not available.

Introduction to Health Emergencies

Health emergencies are unpredictable, often occurring without warning and requiring immediate action to prevent escalation. In any environment, particularly in survival scenarios where access to medical care may be limited or delayed, the ability to manage health emergencies is paramount. This section emphasizes the critical role of medical preparedness, the importance of recognizing common emergencies and the steps necessary to develop an effective emergency plan.

Medical preparedness is the foundation of effective emergency management. It involves more than just having a basic first aid kit; it encompasses the knowledge, skills and mindset required to handle medical crises efficiently and effectively. In survival situations, where professional medical assistance may be hours or even days away, the ability to provide immediate care can be the difference between a minor issue and a life-threatening situation. Preparedness includes understanding how to use medical supplies, knowing the steps to take in various emergencies and being mentally ready to act under pressure.

Preparedness also means being proactive. This involves regularly reviewing and updating your medical supplies, ensuring that they are complete and tailored to the specific risks you might face. For example, if you are in an area prone to snakebites, your kit should include a snakebite kit and antivenom if applicable. Additionally, medical preparedness involves continuous learning, staying informed about the latest first aid techniques and understanding the unique medical challenges presented by your environment, whether it's a remote wilderness area, a flood-prone region or an urban setting.

Mental Health Preparedness

In a survival situation, maintaining mental health is just as crucial as physical health. Stress, anxiety and fear can have significant impacts on decision-making, problem-solving and overall well-being. It's essential to recognize the signs of mental strain and have strategies in place to manage these challenges. Techniques such as deep breathing, mindfulness and establishing a routine can help reduce anxiety and maintain focus. Additionally, maintaining communication with others, even if just within your group, can provide emotional support and reduce feelings of isolation. Remember that taking care of your mental health is not a luxury but a necessity in survival scenarios, ensuring that you can remain calm and clear-headed when making critical decisions.

Identifying Common Emergencies

To prepare effectively for health emergencies, it's crucial to identify the types of emergencies you are most likely to encounter. These can range from minor injuries like cuts and scrapes to more severe situations such as broken bones, burns or allergic reactions. Understanding these common emergencies allows you to tailor your medical supplies and training to address the specific challenges you might face.

For instance, in an outdoor survival scenario, injuries like lacerations, sprains or dehydration are common. On the other hand, in a disaster scenario, such as an earthquake or flood, you might be more concerned with crush injuries, hypothermia or waterborne diseases. By identifying these potential emergencies, you can focus your preparations on the most relevant threats, ensuring that you are ready to respond effectively when they arise.

Creating an Emergency Plan

An emergency plan is a critical component of managing health emergencies. This plan should be comprehensive, covering everything from the roles and responsibilities of each person in your group to the specific steps to take in different types of emergencies. The plan should include detailed instructions for handling various scenarios, such as how to secure a wound, immobilize a fracture or treat a burn, as well as when and how to seek further medical help if available.

Creating an emergency plan also involves planning for contingencies. Consider what you will do if your medical supplies are lost or damaged or if a member of your group suffers from a condition that requires specialized care. Regularly reviewing and practicing your emergency plan ensures that everyone in your group knows what to do, increasing the likelihood of a calm and effective response when a real emergency occurs.

Moreover, your emergency plan should include communication strategies. In a group setting, clear communication is vital to ensure that everyone knows their role and can perform it under stress. Designate a leader who will coordinate the medical response and establish a system for reporting injuries and requesting help. If possible, your plan should also include ways to communicate with external medical services, such as using satellite phones or radio transmitters, in case of severe emergencies.

In summary, being prepared to handle health emergencies is not just about having the right tools, it's about having the right knowledge and a well-practiced plan. By understanding the importance of medical preparedness, identifying common health emergencies and creating a detailed emergency plan, you equip yourself and your group with the ability to respond quickly and effectively to any medical crisis. This preparation is essential not only for survival but also for ensuring the health and well-being of everyone in your care.

Treating Wounds

Treating wounds effectively is a critical skill in managing health emergencies, especially in situations where professional medical assistance is not immediately available. Wounds can range from minor cuts and scrapes to severe lacerations, punctures and burns, each requiring specific treatment to prevent complications such as infection or excessive blood loss. This section explores the various types of wounds, the appropriate treatments for each and essential techniques such as suturing, bandaging and infection prevention.

Types of Wounds and Treatments

Wounds can be categorized into several types, each with distinct characteristics and requiring specific care. The main types include:

- **Abrasions**: these are superficial wounds caused by friction, such as a scrape from a fall. Abrasions typically do not bleed heavily but can be painful and prone to infection. To treat an abrasion, clean the wound thoroughly with clean water to remove any dirt or debris. Apply an antiseptic to prevent infection, then cover the wound with a sterile dressing or bandage.
- **Lacerations**: lacerations are deeper cuts that often involve both the skin and underlying tissues. These wounds can bleed heavily and may require stitches. The first step in treating a laceration is to control the bleeding by applying direct pressure with a clean cloth or bandage. Once the bleeding is under control, clean the wound thoroughly and evaluate whether it needs sutures. If the wound is deep or has jagged edges, suturing might be necessary to close the wound and promote proper healing.
- **Puncture Wounds**: puncture wounds are caused by sharp, pointed objects such as nails or needles. These wounds can be deceptive because they often don't bleed much but can be deep and prone to infection, especially if the object was contaminated. After controlling any bleeding, clean the wound with antiseptic. Do not attempt to close a puncture wound with sutures unless advised by a medical professional, as this can trap bacteria inside the wound, leading to infection.
- **Burns**: burns can be caused by heat, chemicals, electricity or radiation and are classified into first, second and third-degree burns depending on their severity. First-degree burns affect only the outer layer of the skin and can be treated by cooling the burn with running water and applying aloe vera or burn cream. Second-degree burns, which affect deeper layers of the skin and may cause blistering, require more intensive care. After cooling the burn, cover it with a sterile, non-stick dressing and avoid popping any blisters. Third-degree burns are the most severe, damaging all layers of the skin and possibly underlying tissues. These burns require immediate medical attention and while waiting for help, cover the burn with a sterile cloth to protect it from contamination.
- **Avulsions**: an avulsion occurs when a chunk of skin or tissue is torn away from the body. This type of wound often results in heavy bleeding and can be quite serious. To treat an avulsion, first control the bleeding by applying direct pressure. If possible, collect the torn tissue, wrap it in clean gauze and place it in a sealed plastic bag on ice to preserve it for potential reattachment by medical professionals. Cover the wound with a sterile bandage and seek immediate medical help.

Suturing and Bandaging Techniques

Suturing and bandaging are essential techniques for closing wounds, promoting healing and preventing infection. These skills are especially important in situations where medical assistance is not immediately available and the wound is too large or deep to heal on its own without intervention.

Suturing: suturing involves stitching a wound closed to help the skin and tissues heal properly. This technique is typically used for deep cuts or lacerations with jagged edges that cannot be easily bandaged. To suture a wound, you need sterile sutures, a needle, forceps and antiseptic. The process involves cleaning the wound thoroughly, administering local anesthesia if available and stitching the wound closed with even, closely spaced stitches. The goal is to approximate the edges of the wound to allow for proper healing while minimizing the risk of scarring. After suturing, cover the wound with a sterile bandage and monitor it for signs of infection.

Bandaging: bandaging is used to protect a wound, keep it clean and apply pressure to control bleeding. Different types of bandages are used depending on the location and severity of the wound. For example, adhesive bandages are suitable for small cuts and abrasions, while gauze pads and rolls are better for larger wounds. When bandaging a wound, ensure that the dressing is sterile and covers the entire wound. The bandage should be snug enough to stay in place and control bleeding but not so tight that it restricts blood flow.

Preventing Infections

Preventing infection is crucial in wound management, as infections can lead to complications such as abscesses, sepsis or delayed healing. Here are key steps to prevent wound infections:

- **Clean the Wound Thoroughly**: the first step in preventing infection is to clean the wound immediately and thoroughly. Use clean water or a saline solution to rinse the wound and remove any dirt, debris or foreign objects. Avoid using harsh chemicals or alcohol, as they can damage the tissue and delay healing.
- **Apply Antiseptic**: after cleaning the wound, apply an antiseptic solution such as iodine or hydrogen peroxide. This helps kill any bacteria that may be present and reduces the risk of infection. Be sure to follow up with an antibiotic ointment to keep the wound moist and further prevent bacterial growth.
- **Keep the Wound Covered**: covering the wound with a sterile bandage or dressing protects it from external contaminants like dirt, bacteria and moisture. Change the dressing regularly, at least once a day or whenever it becomes wet or dirty. Keeping the wound clean and dry is essential for preventing infection.
- **Monitor for Signs of Infection**: even with proper care, wounds can become infected. Watch for signs of infection, such as increased redness, swelling, warmth, pus or a foul odor from the wound. Fever or increased pain around the wound can also indicate an infection. If any of these signs occur, seek medical attention promptly, as infections can spread quickly and become serious.
- **Use Natural Remedies**: in addition to conventional antiseptics, some natural remedies can aid in preventing infection. Honey, for example, has antibacterial properties and can be used as a topical treatment for wounds. Aloe vera can also soothe the skin and reduce inflammation. However, natural remedies should be used as a complement to, not a replacement for, standard medical care.

Treating wounds effectively involves understanding the type of wound, using appropriate suturing and bandaging techniques and taking proactive steps to prevent infection. By mastering these skills, you can manage wounds effectively in emergencies, reducing the risk of complications and promoting faster healing.

Diseases and Infections

Managing diseases and infections in a survival or emergency situation is a critical aspect of maintaining health and ensuring survival. Without access to modern healthcare, knowing how to identify, treat and prevent common diseases and infections becomes paramount. This section will explore the identification of common diseases, methods for treating infections and the use of natural remedies as part of an overall strategy to maintain health during challenging circumstances.

Identifying Common Diseases

In a survival scenario, some of the most common diseases you may encounter are those related to poor sanitation, contaminated water or close-quarters living. These conditions can lead to the spread of infectious diseases that, if left untreated, can quickly become life-threatening. Common diseases to be aware of include:

- **Respiratory Infections**: diseases such as the common cold, influenza and pneumonia are prevalent in environments where people are in close proximity, especially in damp or cold conditions. Symptoms typically include cough, fever, shortness of breath and fatigue. Without treatment, respiratory infections can escalate into more severe conditions like bronchitis or pneumonia, especially in vulnerable populations such as the elderly or those with compromised immune systems.
- **Gastrointestinal Diseases**: illnesses like diarrhea, dysentery and cholera are often caused by consuming contaminated food or water. Symptoms include stomach cramps, vomiting, diarrhea and dehydration. These diseases can spread rapidly in environments where sanitation is poor, leading to outbreaks that can be difficult to control. Dehydration is a significant risk with gastrointestinal diseases and it can lead to severe complications if not addressed promptly.
- **Skin Infections**: cuts, abrasions and insect bites can lead to skin infections such as cellulitis, impetigo or abscesses if not properly cleaned and treated. Symptoms include redness, swelling, pain and pus. In a survival situation, untreated skin infections can spread and worsen, potentially leading to systemic infections like sepsis.
- **Vector-Borne Diseases**: insects like mosquitoes, ticks and fleas can transmit diseases such as malaria, Lyme disease and dengue fever. These diseases are often characterized by fever, chills, headaches and muscle aches. Preventing bites through the use of insect repellents, protective clothing and nets is crucial in reducing the risk of contracting these diseases.
- **Bloodborne Diseases**: in scenarios where hygiene practices are compromised, bloodborne diseases such as hepatitis B, hepatitis C and HIV can pose a significant risk. These diseases are typically spread through the exchange of blood, often through unsterilized medical equipment or direct contact with infected blood. Symptoms can range from mild flu-like signs to more severe liver damage or immune system failure in the case of HIV.

Treating Infections

Once an infection has been identified, prompt and appropriate treatment is essential to prevent complications. The following are general approaches to treating infections:

- **Antibiotics**: for bacterial infections, antibiotics are the most effective treatment. However, in a survival situation, access to antibiotics may be limited. It is crucial to use antibiotics appropriately and only when necessary to prevent the development of antibiotic resistance. Knowing the symptoms of bacterial infections and having a basic understanding of different types of antibiotics, such as broad-spectrum versus narrow-spectrum, can guide treatment when antibiotics are available.
- **Hydration**: for diseases that cause dehydration, such as those affecting the gastrointestinal system, maintaining hydration is critical. Oral rehydration solutions (ORS) can be prepared using clean water, salt and sugar to replace lost fluids and electrolytes. Ensuring that the patient drinks small, frequent sips can help maintain hydration levels without overwhelming the stomach, which may still be sensitive.
- **Wound Care**: proper care of any cuts, scrapes or other injuries is crucial in preventing infections from taking hold. Clean wounds thoroughly with clean water and apply antiseptics to reduce the risk of infection. Cover the wound with a sterile bandage and change the dressing regularly to keep it clean and dry.
- **Isolation**: in the case of highly contagious diseases, isolating the affected individual can prevent the spread of infection to others. This is particularly important in close-knit environments like survival groups or families where diseases can spread rapidly.
- **Fever Management**: for many infections, fever is a common symptom. While fever is part of the body's natural defense against infection, it can be dangerous if it becomes too high or lasts too long. Using antipyretics like acetaminophen or ibuprofen can help manage fever, but care should be taken not to over-medicate, especially if resources are limited.

Using Natural Remedies

In situations where conventional medical supplies are unavailable, natural remedies can be an effective alternative for managing diseases and infections. Many plants and herbs have medicinal properties that can help treat a variety of conditions:

- **Garlic**: known for its antimicrobial properties, garlic can be used to fight infections. It can be eaten raw, added to food or applied topically to wounds (in a crushed form) to help prevent infection.
- **Echinacea**: this herb is commonly used to boost the immune system and help the body fight off infections. It can be made into a tea or taken as a tincture.
- **Honey**: raw honey has natural antibacterial properties and can be applied directly to wounds to prevent infection. It can also be taken orally to soothe sore throats and help with respiratory infections.
- **Tea Tree Oil**: this essential oil has strong antiseptic and antimicrobial properties. It can be diluted with a carrier oil and applied to the skin to treat infections, insect bites and minor wounds.

- **Ginger**: ginger has anti-inflammatory and antimicrobial effects. It can be used to treat respiratory infections and digestive issues. Ginger tea is particularly effective for soothing nausea and stomach upset.
- **Aloe Vera**: aloe vera is known for its healing properties, especially for skin infections and burns. The gel from the aloe plant can be applied directly to the skin to promote healing and prevent infection.
- **Yarrow**: this plant has been used for centuries to stop bleeding and prevent infection in wounds. It can be applied as a poultice or made into a tea to help with internal infections.

While natural remedies can be effective, it's important to use them knowledgeably and cautiously. Not all natural treatments are safe for everyone and some plants can be toxic if misused. Educating yourself about the proper use of these remedies and their potential interactions with other treatments is crucial for ensuring their safe and effective use.

In summary, managing diseases and infections in a survival scenario requires a combination of knowledge, preparedness and resourcefulness. By understanding the common diseases you may encounter, knowing how to treat infections effectively and utilizing natural remedies when necessary, you can maintain health and prevent the spread of illness in challenging environments. The key is to stay vigilant, act quickly when symptoms appear and adapt your strategies as needed to the resources available to you.

Managing health emergencies in a survival situation requires a combination of preparation, knowledge and practical skills. By understanding the importance of medical preparedness, mastering wound treatment and being able to identify and treat diseases and infections, you can significantly increase your chances of surviving and thriving in challenging conditions. Remember, the key to successful health management is not just having the right tools but knowing how to use them effectively. Stay prepared, stay informed and always be ready to act swiftly in an emergency to protect your health and the health of those around you.

Exercise Chapter 19
Practicing First Aid Techniques with Family

Objective: ensure that all family members are familiar with basic first aid techniques and can confidently respond to common health emergencies.

Materials Needed: first aid kit, sterile bandages, antiseptic wipes, adhesive tape, sterile gauze pads, tweezers, scissors, a practice dummy or soft pillow and a printed first aid manual.

1. Review First Aid Basics: begin by gathering your family members and reviewing the contents of your first aid kit. Explain the purpose of each item, such as bandages, antiseptic wipes and gauze. Discuss common emergencies like cuts, burns and sprains and explain the appropriate first aid response for each.

2. Practice Treating Wounds: using the practice dummy or a pillow, demonstrate how to clean and bandage different types of wounds. Show how to apply pressure to stop bleeding, how to clean a wound with antiseptic wipes and how to securely bandage it. Have each family member take turns practicing these techniques to ensure they understand how to handle real-life scenarios.

3. Learn and Practice Suturing Techniques: if you have access to a suture practice kit, teach your family the basics of suturing. Explain when and why suturing is necessary and let each person practice the technique on the kit or a similar substitute. Make sure to emphasize the importance of cleanliness and precision during this process.

4. Simulate a Health Emergency: create a realistic scenario where one family member "pretends" to have an injury or illness. The rest of the family must respond appropriately, using the first aid techniques reviewed earlier. After the simulation, discuss what went well and what could be improved in a real emergency.

Deliverable: write a brief report summarizing your family's first aid practice session. Include details on each family member's performance, any challenges encountered and how they were addressed. Attach photos of the practice scenarios, especially of the simulated health emergency. Reflect on the experience and suggest areas for improvement in future practice sessions.

Practicing first aid techniques as a family not only equips everyone with the necessary skills to respond to health emergencies but also builds confidence and teamwork. Regular practice sessions ensure that when an emergency arises, everyone knows their role and can act swiftly and effectively to protect the health and safety of the family.

Chapter 20
Managing Community Resources

In any survival situation, managing community resources is critical to ensuring the well-being and sustainability of the group. While individual preparedness is important, there is strength in numbers and a well-organized community can achieve much more than isolated individuals. When resources are pooled and managed effectively, communities can survive and even thrive in difficult circumstances. This chapter explores the principles of community management, the importance of teamwork and mutual support and strategies for effectively managing and sharing resources within a group.

Effective community management begins with understanding the unique dynamics of group cohesion and the essential role it plays in crisis situations. A community that works together, communicates effectively and supports one another is far more resilient than a group of individuals working independently. However, managing a community is not without its challenges. Differences in opinions, conflicting priorities and the stress of a survival scenario can strain relationships and threaten the cohesion of the group. Therefore, it's essential to establish clear roles, open communication channels and a strong sense of shared purpose from the outset.

This chapter is divided into three sections. The first section discusses the importance of community cohesion, mutual support and the differences between managing resources individually and within a group. The second section focuses on building an effective community, emphasizing the importance of teamwork, creating support networks and maintaining effective communication in crisis situations. The final section addresses the management and sharing of resources, highlighting the importance of collaboration, the sharing of knowledge and skills and the role of community support and mutual aid in ensuring survival.

Introduction to Community Management

Managing community resources, especially in the context of survival or crisis scenarios, is a task that extends far beyond the logistics of distributing supplies and assigning roles. It is about creating and sustaining a unified group, where cooperation and mutual support are paramount to overcoming challenges and ensuring the well-being of all members. Community management is deeply rooted in the principles of leadership, psychology and social dynamics, requiring a nuanced approach that balances individual needs with the collective good.

At the core of successful community management is group cohesion, the glue that binds members together. Group cohesion is not just a matter of getting along; it's the foundation upon which trust, cooperation and shared purpose are built. A cohesive group is more likely to work together effectively, share resources fairly and support one another in times of need. This cohesion is fostered through clear communication, mutual respect and the establishment of common goals. Leaders play a crucial role in maintaining this cohesion by facilitating open dialogue, addressing conflicts promptly and ensuring that every member feels valued and heard.

In any community, especially during crises, the concept of mutual support becomes vital. Mutual support refers to the reciprocal nature of help and assistance within a group. It's the understanding that everyone contributes to the community's well-being and in return, they can rely on others for support when needed. This support network is essential in maintaining morale and resilience during difficult times. It's not just about sharing physical resources, but also about providing emotional and psychological support, which can be just as critical in ensuring a group's survival.

Another essential aspect of community management is understanding the differences between individual and group management. While managing resources for oneself involves straightforward decision-making focused on personal survival, managing a community requires a broader perspective. In a group setting, decisions must consider the needs of all members, which can complicate the process. The shift from self-reliance to interdependence is crucial, where the well-being of the group takes precedence over individual desires. This requires a mindset change, where cooperation, compromise and collective responsibility become the guiding principles.

Effective community management also involves collaborative decision-making. Unlike individual decision-making, which can be swift and unilateral, group decisions often involve multiple perspectives and require consensus-building. This process can be time-consuming, but it's necessary for ensuring that all voices are heard and that decisions are made with the group's best interest in mind. Collaborative decision-making fosters a sense of ownership among community members, making them more committed to the group's success.

Furthermore, managing a community involves the allocation and sharing of resources. In a survival scenario, resources are often limited and how they are distributed can impact the group's cohesion and survival. Effective resource management involves prioritizing needs, rationing supplies and ensuring that no member is left without essential provisions. This requires transparency, fairness and sometimes difficult decisions about who gets what and when.

Leadership plays a pivotal role in all these aspects of community management. A strong leader is not just someone who gives orders but is someone who inspires, guides and unites the group. Leadership in a community setting is about serving the group's interests, mediating conflicts and ensuring that everyone is working towards the same goals. It's about being adaptable, empathetic and decisive when necessary.

In summary, community management is a multifaceted and dynamic process that requires more than just organizational skills. It's about building a strong, cohesive group where mutual support, collaborative decision-making and effective resource management are prioritized. By fostering these elements, a community can become a resilient and unified force, capable of overcoming the challenges that crises inevitably bring. The success of a community in such scenarios hinges on its ability to work together, support each other and manage resources wisely, ensuring that all members survive and thrive.

Building an Effective Community

Building an effective community, particularly in a survival or crisis scenario, is a complex process that requires more than just gathering individuals together. It involves creating a cohesive unit where teamwork, mutual support and effective communication are central to the community's functioning and success. This section explores the key components of building a strong and resilient community, focusing on the importance of teamwork, the creation of support networks and the necessity of clear communication, especially during crises.

Importance of Teamwork

Teamwork is the cornerstone of any successful community. It is the mechanism by which individuals with diverse skills, experiences and resources come together to achieve common goals. In a survival scenario, the collective effort of a group often determines whether they thrive or fail. Effective teamwork requires that members of the community understand their roles, trust one another and work collaboratively towards shared objectives.

To foster teamwork, it's crucial to establish a culture of cooperation from the outset. This involves setting clear goals for the community and ensuring that every member understands how their contributions fit into the bigger picture. Leaders play a vital role in facilitating teamwork by creating opportunities for collaboration, resolving conflicts quickly and encouraging an environment where every member's input is valued. Trust-building activities, regular group meetings and collaborative problem-solving exercises can strengthen the bonds between community members, making them more likely to support each other in times of need.

Furthermore, it's important to recognize and utilize the diverse skills and strengths of each community member. Whether someone excels in medical care, resource management, communication or physical labor, allowing them to contribute in areas where they are most skilled not only boosts the efficiency of the group but also increases individual satisfaction and engagement.

Creating Support Networks

A support network within a community is essential for both practical and psychological reasons. Practically, a support network ensures that resources, skills and assistance are shared effectively, allowing the community to function smoothly and respond to challenges. Psychologically, knowing that others are there to help in times of need strengthens individual resilience and community cohesion.

Support networks are built on the principles of reciprocity and mutual aid. Reciprocity involves the exchange of help or resources with the expectation that assistance will be given in return when needed. This fosters a sense of balance and fairness within the community, as everyone contributes and benefits in turn. Mutual aid extends this concept by emphasizing the collective responsibility of the community to look after its members, particularly those who are most vulnerable, such as the elderly, children or those with health issues.

To create effective support networks, it's essential to identify the needs and resources of each community member. Regular check-ins and open communication can help ensure that everyone's needs are met and that no one is left behind. Additionally, establishing a system of resource sharing – whether it's food, medical supplies or knowledge – can help ensure that all members have access to what they need.

In crisis situations, support networks are vital for maintaining morale and providing the necessary care and resources to those affected. Whether it's organizing a group to help rebuild after a disaster, providing emotional support during stressful times or pooling resources to address a common threat, these networks are the backbone of a resilient community.

Effective Communication in Crisis Situations

Communication is the lifeblood of a well-functioning community, especially in crisis situations where clear, accurate and timely information can mean the difference between life and death. Effective communication ensures that everyone is informed about the community's goals, plans and any changes in the situation. It also helps to prevent misunderstandings and conflicts, which can undermine the community's cohesion and effectiveness.

In building an effective community, it's crucial to establish clear communication channels that everyone can access. This might involve regular meetings, written notices or digital communication tools, depending on the resources available. It's also important to establish protocols for communication during emergencies, such as who will be responsible for disseminating information, how decisions will be communicated and what methods will be used if primary communication systems fail.

Crisis situations often require rapid decision-making and the ability to convey instructions clearly and concisely. Leaders must be able to communicate not just what needs to be done, but why it's important, to ensure buy-in from all members. In times of high stress, maintaining a calm, clear and authoritative communication style can help keep the community focused and prevent panic.

Moreover, communication is not just about giving orders, it's also about listening. In a community, every member's voice should be heard and their concerns and suggestions should be taken into account. Encouraging open dialogue and creating a culture where members feel safe to express their thoughts can lead to better decision-making and stronger group cohesion.

Building an effective community is about creating a supportive, communicative and collaborative environment where every member plays a role in the group's success. By emphasizing teamwork, developing robust support networks and ensuring effective communication, a community can become a resilient unit capable of facing any challenge together. These elements are not just important for surviving crises, but for thriving as a cohesive group in any situation.

Managing and Sharing Resources

Managing and sharing resources within a community, especially in times of crisis, is a critical aspect of ensuring collective survival and well-being. In such situations, resources can be scarce and the ability to effectively manage what is available, distribute it fairly and utilize everyone's skills and knowledge becomes essential. This section will explore the importance of collaboration, the benefits of sharing knowledge and skills and the role of community support and mutual aid in resource management.

Importance of Collaboration

Collaboration is at the heart of successful resource management in any community. When individuals come together to pool their resources — whether those are physical goods, skills or knowledge — the collective strength of the community is greatly enhanced. Collaboration allows for the efficient use of resources, reduces waste and ensures that everyone's needs are met as much as possible.

In crisis situations, where resources such as food, water, medical supplies and shelter may be limited, collaboration becomes even more crucial. By working together, community members can create systems of shared responsibility and mutual support. This might involve forming teams to manage different aspects of the community's needs, such as food distribution, medical care or security. Collaborative efforts also help to prevent the concentration of resources in the hands of a few, promoting fairness and equity within the group.

Effective collaboration requires clear communication and a shared understanding of the community's goals. It also involves recognizing and valuing the contributions of each member, regardless of the size or nature of their contribution. Trust is a key component in this process, as individuals need to believe that their efforts will be reciprocated and that the group will work together to solve problems and overcome challenges.

Sharing Knowledge and Skills

One of the most valuable resources in any community is the collective knowledge and skills of its members. Sharing this knowledge is crucial for the survival and growth of the community. In a survival scenario, skills such as first aid, food preservation, construction and navigation are invaluable. The more these skills are shared and taught within the community, the better prepared the group will be to handle various challenges.

Knowledge sharing can take many forms, from formal training sessions and workshops to informal mentoring and daily collaboration. For example, someone with experience in agriculture might teach others how to grow and preserve food, while a person with medical training could conduct first aid workshops. This exchange of knowledge not only enhances the skills of all community members but also fosters a sense of unity and shared purpose.

In addition to practical skills, sharing knowledge about the environment, local resources and potential risks is vital. Understanding the local ecosystem, weather patterns and available resources can help the community plan more effectively and make informed decisions about resource management. This kind of knowledge is often specific to the local area and can be crucial in adapting to changing conditions or emergencies.

Furthermore, the act of teaching and learning within the community helps to build relationships and trust. When members feel that their contributions are valued and that they are learning from others, they are more likely to be engaged and committed to the group's success. This spirit of cooperation and mutual aid is essential for maintaining morale and ensuring long-term resilience.

Community Support and Mutual Aid

At the core of managing and sharing resources is the principle of mutual aid. Mutual aid involves the voluntary reciprocal exchange of resources and services for mutual benefit. In a community setting, this means that members support each other by sharing what they have, whether it's food, tools, labor or expertise, with the understanding that others will do the same when needed.

Mutual aid is not a formal system of charity, but rather a horizontal form of support where everyone is both a giver and a receiver. This creates a strong sense of solidarity and reduces the power imbalances that can arise in hierarchical systems of aid. In a survival situation, mutual aid ensures that resources are distributed according to need rather than status or wealth, helping to maintain equity and fairness within the community.

Community support goes beyond just sharing physical resources. Emotional and psychological support are equally important, particularly in crisis situations where stress and anxiety can run high. By fostering a supportive environment where members look out for one another, the community can build resilience against the psychological toll of prolonged crises.

To organize mutual aid effectively, communities often establish systems for tracking resources, identifying needs and coordinating efforts. This might involve creating a community inventory of available resources, setting up distribution centers or forming committees to oversee different aspects of resource management. These systems help to ensure that aid is delivered where it's needed most and that the community can respond flexibly to changing circumstances.

In conclusion, managing and sharing resources within a community requires a combination of collaboration, knowledge sharing and mutual aid. By working together, sharing what they know and have and supporting one another, community members can create a resilient and adaptable group capable of facing any challenge. These principles not only help ensure survival in times of crisis but also strengthen the bonds between individuals, fostering a sense of collective responsibility and solidarity that benefits everyone.

Exercise Chapter 20
Establishing a Community Support Network

━━━ ✧✦✧ ━━━

Objective: develop a community support network that facilitates mutual aid, resource sharing and effective communication among members.

Materials Needed: pen and paper or a digital note-taking tool, a group of community members, a meeting space (physical or virtual), communication tools (phones or radios) and a whiteboard or chart paper for brainstorming.

1. Identify Key Members: gather a group of individuals committed to building a community support network. Ensure members represent various skills, resources and perspectives. Assign roles like coordinator, communication lead and resource manager to distribute tasks efficiently.

2. Map Out Resources: conduct a survey or brainstorming session to identify available resources within the community, including food supplies, medical kits, tools and skills. Create a resource map detailing access points and how these resources can be shared or distributed when needed.

3. Establish Communication Channels: decide on methods for keeping everyone informed and connected, such as a phone tree, group chat or radios. Ensure all members know how to use these tools and establish protocols for regular check-ins, especially during emergencies.

4. Develop a Mutual Aid Agreement: create a simple agreement outlining how members will support each other in times of need, including resource sharing, requesting help and participation expectations. Ensure the agreement is clear, fair and agreed upon by all members.

5. Practice and Revise: schedule regular meetings or drills to practice the support network's response to different scenarios. Use these sessions to identify weaknesses or areas for improvement and adjust the plan based on group feedback.

Deliverable: write a brief report detailing the establishment of your community support network, including roles assigned, resources identified and communication channels set up. Attach any agreements made and reflect on initial practice sessions. Discuss any challenges encountered and how they were addressed.

Establishing a community support network strengthens the group's ability to manage resources and provides a safety net for all members. By building this network, you not only enhance the community's resilience but also foster a spirit of cooperation and mutual aid that benefits everyone involved.

Conclusion
Closing the work

As we draw this journey to a close, it's essential to reflect on the path we've taken and the lessons we've learned. This book has been more than a guide; it has been a roadmap to self-sufficiency, a call to reclaim skills that our ancestors once mastered and a challenge to adapt those skills to our modern world. In a time where uncertainty often looms large, the knowledge and practices we've explored are not just about survival but about thriving in the face of adversity.

Whether you are preparing for a potential crisis, aiming to live a more self-reliant life or simply interested in acquiring practical skills, this journey equips you to face the future with confidence and resilience.

Summary of Key Lessons

Throughout this book, we've embarked on a comprehensive journey through the many facets of survival, self-sufficiency and crisis management. From the psychological resilience needed in survival situations to the practical skills required for collecting water, growing food and ensuring personal security, the lessons outlined in these chapters form a holistic guide to thriving in uncertain times.

Mental Preparedness and Psychological Resilience: one of the foundational elements we've explored is the importance of mental preparation. Survival isn't just about physical endurance, it's also about cultivating a mindset that can withstand the stresses and anxieties that arise in crisis situations. We delved into techniques for managing stress, staying calm under pressure and fostering psychological resilience. Understanding group dynamics, managing relationships and building strong communication within a community were also emphasized as critical components of a cohesive survival strategy.

Essential Survival Skills: practical skills form the bedrock of survival. We've covered the essentials, from water collection and purification to food production and preservation. Whether it's harvesting rainwater, growing your own vegetables or preserving food through drying, smoking or canning, these skills ensure that you can maintain a steady supply of essentials even when traditional resources are unavailable. Additionally, we discussed the construction of alternative energy systems, like solar panels and wind generators, which provide the means to sustain modern conveniences in off-grid scenarios.

Building and Securing Your Environment: creating a safe and secure living environment is paramount in any survival situation. Chapters on home security, emergency shelter construction and the development of emergency structures provided step-by-step guidance on how to protect your property and ensure your family's safety. Whether you're building a permanent shelter or setting up a temporary one in the wilderness, the techniques discussed are designed to maximize both security and comfort.

Health and Medical Preparedness: the ability to manage health emergencies is crucial when access to professional medical help is limited. We explored the creation of first aid kits, the treatment of wounds and the management of common diseases and infections using both conventional and natural remedies. By understanding how to identify and treat health issues promptly, you can prevent minor injuries from becoming life-threatening situations.

Community and Resource Management: as we moved into more advanced topics, the focus shifted to managing community resources and fostering mutual aid. We discussed how to build effective communities, emphasizing the importance of teamwork, communication and resource sharing. The lessons here are not just about survival, they're about creating resilient communities that can thrive through collaboration and shared purpose.

Navigation and Disaster Preparedness: navigation and orientation techniques were covered to ensure you can find your way in unfamiliar terrains, whether by using a compass and map or by relying on natural signs like the sun and stars. Disaster preparedness was another key area, where we explored how to plan for various natural disasters, create emergency kits and develop family plans to ensure everyone knows what to do in a crisis.

In summary, the key lessons from this book are not just about learning specific skills or strategies, but about adopting a mindset of continuous preparation and adaptability. Whether you're facing the daily challenges of self-sufficiency or the sudden impact of a disaster, these lessons provide the tools and knowledge needed to face any situation with confidence and resilience.

Importance of Continuous Preparation

The journey toward self-sufficiency and survival readiness does not end with the completion of a single project or the mastery of a particular skill. Instead, it is an ongoing process that requires continuous preparation, learning and adaptation. This section underscores why continuous preparation is crucial for maintaining and enhancing your ability to survive and thrive in various situations.

Adapting to Changing Circumstances: the world around us is constantly changing, environmental conditions shift, new threats emerge and personal circumstances evolve. Continuous preparation allows you to adapt to these changes effectively. Whether it's a shift in the climate that affects your food production, a new technology that enhances your energy efficiency or a change in your community's structure, staying prepared means staying flexible. Regularly updating your skills, knowledge and resources ensures that you remain ready for whatever challenges may arise.

Maintaining Skills and Equipment: survival skills, like any other abilities, can degrade over time if not practiced regularly. Just as a muscle weakens without exercise, your proficiency in crucial survival tasks – such as fire-starting, first aid or navigation – can diminish without ongoing practice. Continuous preparation involves not just learning new skills, but also maintaining and refining the ones you already have. This also applies to your equipment: tools, emergency kits and other resources must be regularly checked, maintained and replaced as needed to ensure they are in optimal condition when required.

Building a Resilient Mindset: continuous preparation fosters a resilient mindset that is crucial in survival situations. This mindset is characterized by the ability to remain calm under pressure, adapt to new circumstances and find solutions in the face of adversity. By continually preparing, you train your mind to expect and handle challenges, reducing the likelihood of panic or despair when faced with a crisis. Regularly engaging in problem-solving activities, drills and simulations helps to build this mental resilience, ensuring that you can think clearly and act decisively in emergencies.

Enhancing Community Strength: in the context of community survival, continuous preparation is not just an individual responsibility but a collective one. Communities that regularly engage in preparedness activities, such as group training sessions, resource assessments and emergency drills, are more likely to remain cohesive and effective in a crisis. Continuous preparation helps to strengthen the bonds between community members, fostering a sense of trust and mutual support that is vital for collective resilience. By staying prepared together, communities can better manage shared resources, coordinate their efforts and ensure the well-being of all members.

Staying Ahead of Potential Threats: the world is full of potential threats, from natural disasters and economic crises to pandemics and social unrest. Continuous preparation allows you to stay ahead of these threats by keeping you informed and ready to respond. This proactive approach involves monitoring current events, staying updated on best practices in survival and self-sufficiency and regularly revising your emergency plans to address new risks. By anticipating potential challenges and preparing for them in advance, you can minimize their impact and increase your chances of survival.

Promoting Long-Term Sustainability: ultimately, continuous preparation is about promoting long-term sustainability in your lifestyle and survival strategies. It's about ensuring that the systems you've put in place — whether they're for food production, energy generation or community management — can withstand the test of time and adapt to future challenges. By continuously preparing, you invest in the long-term resilience of your environment, your community and yourself, ensuring that you are not just surviving, but thriving, in the face of adversity.

Continuous preparation is the key to maintaining readiness, enhancing resilience and ensuring long-term survival. It's a commitment to lifelong learning, regular practice and proactive adaptation that keeps you prepared for whatever the future may hold. By embracing this approach, you equip yourself with the tools, knowledge and mindset necessary to navigate an uncertain world with confidence and competence.

Future Perspectives on Self-Sufficiency

As we look ahead, the concept of self-sufficiency is likely to evolve, driven by advances in technology, changes in global socio-economic conditions and an increasing awareness of environmental sustainability. This section explores the future of self-sufficiency, considering how emerging trends and innovations might shape the way individuals and communities approach self-reliance in the years to come.

Technological Advancements in Self-Sufficiency: one of the most significant factors influencing the future of self-sufficiency is the rapid pace of technological innovation. From renewable energy systems to smart farming techniques, technology is making it easier than ever to achieve a high degree of self-reliance. Advances in solar, wind and hydroelectric power generation are enabling individuals and communities to produce clean, sustainable energy on a small scale, reducing dependence on centralized power grids. Innovations in water purification and waste management technologies are also empowering people to manage their resources more efficiently, even in off-grid or remote locations.

In agriculture, the rise of precision farming, hydroponics and vertical gardening is revolutionizing how we grow food. These technologies allow for more efficient use of space, water and nutrients, making it possible to produce food in urban environments or areas with limited arable land. Additionally, the development of AI and robotics in agriculture and home maintenance is likely to further enhance self-sufficiency by automating routine tasks and optimizing resource management.

The Role of Community and Collaboration: while individual self-sufficiency will continue to be important, the future is likely to see a greater emphasis on community-based self-reliance. As the challenges we face — such as climate change, economic instability and social unrest — become more complex, the need for collective action and mutual support will grow. Communities that work together to share resources, skills and knowledge will be better positioned to withstand these challenges and create resilient, sustainable systems.

The concept of localism, which promotes the use of local resources and supports local economies, is likely to play a significant role in this shift. By fostering strong, interconnected communities that prioritize local production and consumption, we can reduce our dependence on global supply chains and create more resilient, self-sufficient societies. This trend toward localism is already evident in the growing popularity of farmers' markets, community-supported agriculture (CSA) programs and local energy cooperatives.

Sustainability and Environmental Stewardship: as concerns about environmental degradation and climate change continue to rise, the future of self-sufficiency will be increasingly tied to sustainability. Self-reliance practices will need to incorporate principles of environmental stewardship, ensuring that the resources we depend on are managed in a way that preserves them for future generations. This means adopting sustainable farming practices, reducing waste, conserving water and minimizing our carbon footprint.

Permaculture, a design philosophy that emphasizes working with nature rather than against it, is likely to gain prominence as a framework for sustainable self-sufficiency. Permaculture principles – such as creating closed-loop systems, promoting biodiversity and using renewable resources – offer practical solutions for living sustainably while maintaining a high degree of self-reliance. As more people recognize the importance of living in harmony with the environment, permaculture and other sustainable practices will become central to the future of self-sufficiency.

Adaptation to Global Uncertainties: the future is uncertain and self-sufficiency will play a critical role in helping individuals and communities adapt to whatever challenges lie ahead. Whether it's dealing with the effects of climate change, navigating economic disruptions or responding to global pandemics, the ability to be self-reliant provides a safety net that enhances security and resilience.

Preparedness will continue to be a key aspect of self-sufficiency in the future. This involves not only stockpiling resources and developing survival skills but also staying informed about global trends and potential risks. As the world becomes more interconnected, the ability to anticipate and adapt to changes will be crucial for maintaining self-reliance. This might involve diversifying your skill set, building strong community networks or investing in technologies that enhance your ability to live independently.

Empowerment Through Knowledge and Skills: ultimately, the future of self-sufficiency is about empowerment, empowering individuals to take control of their lives, reduce their reliance on external systems and build a future that aligns with their values. Knowledge and skills will be at the heart of this empowerment, providing the tools needed to navigate an increasingly complex world.

Education and skill development will therefore be critical in promoting self-sufficiency. As traditional education systems evolve, there may be a growing emphasis on teaching practical skills – such as gardening, carpentry, renewable energy and food preservation – that are essential for self-reliance. Online platforms, workshops and community learning initiatives will play a key role in disseminating this knowledge, making it accessible to people from all walks of life.

The future of self-sufficiency is rich with promise and potential. By embracing technological advancements, fostering strong community ties and committing to sustainable practices, we create not just a survival strategy, but a way of life that thrives in harmony with the challenges ahead. The journey toward self-reliance is continuous, evolving with each new skill learned and each resource managed. Together, with the right mindset and tools, we can build a resilient, sustainable world where self-sufficiency is not just a necessity but a fulfilling reality.

Embarking on the path of self-sufficiency is a powerful act of taking control of your future. Every step you take, every skill you acquire, strengthens your autonomy and secures your place in an uncertain world. This journey is about more than survival; it's about building a life of resilience, security and fulfillment. Whether you're at the beginning or well along your way, remember that this path leads not just to independence but to a richer, more empowered life for you and your community. Embrace this journey with confidence, knowing that each challenge met is a victory for your self-reliance and well-being.

Bonus
Your Exclusive Survival Bonuses

Welcome to the bonus page!
Here you can download your survival bonuses:

1. **"No Grid Survival Essentials – 7 Printable Cheat Sheets for Self-Sufficiency"**, a collection of seven printable guides designed to enhance your self-sufficiency in real-world scenarios.
2. **"The 72-Hour Survival Blueprint"**, your step-by-step guide to making the right decisions in the most critical moments, staying ahead of danger and securing your long-term survival.
3. **"Minimalist Survival":** this guide reveals the core principles of surviving with minimal resources, maximizing efficiency, minimizing waste and thriving off-grid using only the bare essentials.
4. **"Baofeng Radio Bible - Survival Handbook"**, a comprehensive 120-page ebook for mastering Baofeng radio communications, ensuring reliable survival communication and managing emergency situations.

SCAN THIS QR CODE TO DOWNLOAD YOUR BONUSES INSTANTLY!

About Author
Some words about the writer

Alexander Freeman hails from a quaint, rural town nestled in the picturesque landscapes of Montana, where the sprawling countryside and the community's close-knit fabric laid the groundwork for his lifelong passions and expertise. From a young age, Alexander was captivated by the workings of the world around him, often found dismantling gadgets only to reassemble them or out in the fields exploring the intricacies of nature. This curiosity for the practical aspects of life evolved into a profound expertise in technology, encompassing everything from the complexities of computer systems to the nuanced world of radio communications.

Alexander's journey into the realm of technological mastery was not just about accumulating knowledge; it was also about embracing a lifestyle that values self-reliance and independence. His approach to life is a testament to the belief that one can achieve a harmonious balance with technology while also fostering self-sufficiency. This philosophy extends beyond just understanding how devices work; it encompasses a broader spectrum of autonomy, including energy independence, technological self-reliance and even growing one's own food.

As an author, Alexander Freeman brings this rich tapestry of experiences and beliefs to the pages he writes, aiming to empower his readers with the knowledge and skills to navigate the modern world while maintaining a strong sense of self-sufficiency and independence. His works are more than just guides; they are a reflection of a life lived with purpose and a deep connection to the technological and natural worlds. Through his writing, Alexander hopes to inspire others to explore the potential of technology not as a crutch but as a tool for building a more autonomous and fulfilling life.

Good luck with everything!

Table of Contents

―――― ✧✦✧ ――――

Introduction	3
Chapter 1	8
Chapter 2	16
Chapter 3	23
Chapter 4	30
Chapter 5	38
Chapter 6	46
Chapter 7	54
Chapter 8	62
Chapter 9	71
Chapter 10	80
Chapter 11	87
Chapter 12	96
Chapter 13	107
Chapter 14	116
Chapter 15	124
Chapter 16	133
Chapter 17	141
Chapter 18	150
Chapter 19	158
Chapter 20	167
Conclusion	175
Bonus	181
About Author	182

Made in the USA
Columbia, SC
16 March 2025

54959695R00100